SIDA, BOTANICAL MISCELLANY 11
BOTANICAL RESEARCH INSTITUTE OF TEXAS, INC.

TEXT ANNOTATIONS AND IDENTIFICATION NOTES FOR

MANUAL OF THE VASCULAR FLORA OF THE CAROLINAS

JOHN D. FREEMAN
Department of Botany and Microbiology
Auburn University, AL 36849-5407, USA

1994

ISSN 0833-1475

SIDA, BOTANICAL MISCELLANY NO. 11
FOUNDED BY
WM. F. MAHLER AND BARNEY L. LIPSCOMB, 1987
EDITOR: BARNEY L. LIPSCOMB
DESIGNER: LINNY HEAGY
DIRECTOR: S.H. SOHMER
BOTANICAL RESEARCH INSTITUTE OF TEXAS, INC.
509 PECAN STREET
FORT WORTH, TEXAS 76102-4060 USA

COVER PHOTO:
VENUS' FLY TRAP, ***DIONAEA MUSCIPULA*** *ELLIS, A MONOTYPIC GENUS AND FAMILY NATIVE ONLY TO THE COASTAL PLAIN OF THE CAROLINAS. THIS SPECIES HAS BEEN INTRODUCED INTO BOGS AND SAVANNAHS IN NORTHERN FLORIDA AND SOUTHERN ALABAMA WHERE POPULATIONS HAVE BECOME ESTABLISHED AND APPEAR TO BE REPRODUCING BY SEED.*
PHOTO BY DEBBIE FOLKERTS

FRONTISPIECE:
LINE ILLUSTRATION OF ***TRILLIUM PUSILLUM*** *MICHAUX, DESCRIBED IN 1803 ON THE BASIS OF COLLECTIONS FROM THE LOW COUNTRY OF SOUTH CAROLINA NEAR CHARLESTON. ONLY A SINGLE POPULATION IS PRESENTLY KNOWN TO REMAIN IN SOUTH CAROLINA.*
ILLUSTRATION BY JOHN D. FREEMAN

BRIT

Table of Contents

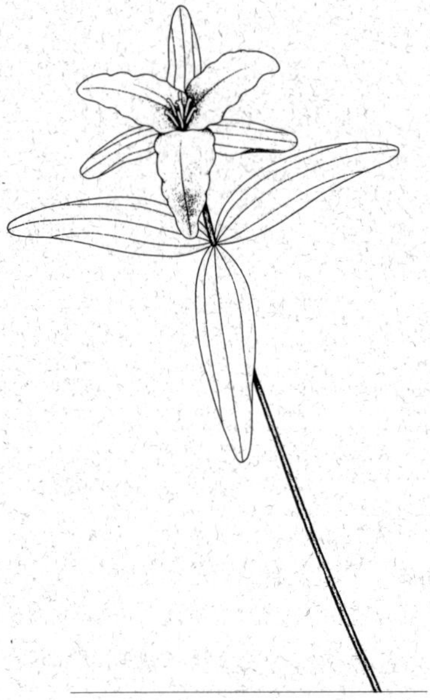

Trillium pusillum Michaux

IV

Text Annotations and Identification Notes for

Manual of the Vascular Flora of the Carolinas

John D. Freeman
Department of Botany and Microbiology
Auburn University, AL 36849-5407, U.S.A.

Abstract

Six hundred-seven annotations and notes to aid plant identification are provided for the *Manual of the Vascular Flora of the Carolinas* by Radford, Ahles, and Bell. These include 265 concerning inconsistencies between keys and descriptions, 215 that correct spelling and other simple printing errors, 94 glossary additions and clarifications of terms, and 33 that deal either with nomenclature or authorship for accepted names.

Resumen

Se ofrecen seiscientas siete anotaciones y notas de ayuda a la identificación para el *Manual of the Vascular Flora of the Carolinas* de Radford, Ahles y Bell. De ellas, 265 son relativas a inconsistencias entre claves y descripciones, 215 para corregir orthografía y errores tipográficos, 94 adiciones al glosario y clarificaciones de términos, y 33 que tratan tanto de la nomenclatura como de la autoría de los nombres aceptados.

During the past quarter century, *Manual of the Vascular Flora of the Carolinas* (Radford, Ahles, and Bell 1968) has become *the* standard reference for plant identification throughout the Southeast. It has served very well as a lab manual and teaching aid for introductory plant taxonomy and field botany courses from Virginia to Florida and westward into Louisiana and Arkansas and consequently has become the most widely used flora in the region. Some of this acceptance has resulted because Small's *Manual* (1933) was already nomenclaturally obsolete before work on the Carolinas *Manual* was begun. Lack of a recent regional manual has contributed somewhat, but the main reason for its popularity is that (except for peninsular Florida) it serves as a fairly reliable source of names for much of the southeastern flora outside the Carolinas. In comparison with other references, especially the new *Vascular Flora of the Southeastern United States* volumes now available (Cronquist 1980, Isely 1990), it remains relatively inexpensive. Unfortunately, although it has gone through more than 11 printings (that is, 11 *numbered* ones through 1988, some of which included minor corrections), most of the original printing errors and inconsistencies between keys and descriptions continue to appear in the text.

Some time ago, in connection with use of this manual in plant taxonomy courses at Auburn University, I started a card file to keep a cumulative record of simple spelling errors and other problems discovered by students as they attempted to identify unknown specimens. A decline in the number of new problems reported during recent quarters led me to conclude that few serious mistakes remain to be found in the *Manual*. Lists of annotations and corrections for two other floras (Correll 1972, Hill 1981, Johnston 1990) evidently have proven useful. The probability that the same would be true for the Carolinas *Manual* led me to distribute the list in manuscript form to my classes. I am grateful to more than 200 former students, other colleagues, and anonymous contributors (including persons who reviewed versions of this list) for detecting many of the problems that are treated. Those just getting acquainted with "R.A.B." or "RAB" (as the students call it) can now correct their copies before needlessly experiencing some of the same frustrations as have others before them.

Annotations and identification notes for *Manual of the Vascular Flora of the Carolinas* are treated herein as one continuous listing. All entries are according to page number. Spelling and wording errors and inconsistencies between keys and descriptions, as stated above, served as the basis for the first list that was used in my classes, and any sort of distracting or misleading entry (480 in all) was regarded as one that ought to be fixed by means of some change in wording or punctuation. This is because a minor flaw in something as technical as a manual can discourage beginning students when such problems can be managed more easily by experienced taxonomists. Only undotted i's, several of which occur in the *Manual*, were regarded as trivial enough to be excluded from this list. Corrections were provided for most spelling errors, but some minor inconsistencies were merely noted, whether erroneous or not, without any change being suggested. Problems with glossary terms were targeted for clarification: a total of 94 major changes and definitions were added for terms used in keys and descriptions but either omitted from the *Manual* glossary or incorrectly treated in it.

Apparent differences in specimens collected outside the Carolinas (especially from Alabama southward and westward) were common among the problems that were first noted concerning the RAB *Manual*. Morphological features that seemed to be different in non-Carolina plants from those described for the Carolinas flora might be due either to geographical variation or some error in the manual; we were not able to determine the exact cause for these discrepancies in many cases, and Radford, Ahles, and Bell should not be held accountable for morphological features or geographical ranges of plants growing outside the area for which the *Manual* was specifically written. Annotations concerning the morphology of collections made outside the Carolinas should be viewed simply as notes and not necessarily corrections.

Thirty-three (33) nomenclatural changes and corrections in the authorship for plant names used in the *Manual* complete this list; these changes are mainly for plants with which I have either some research experience firsthand or a bias that required expression.

An up-to-date nomenclatural revision was not an objective in compiling the list, so it is very imperfect and incomplete in this regard. We can expect nomenclature to be continually revised and changed for plant taxa as long as research is being done. Other sources (Kartesz 1994, Morin 1993) may help to resolve many matters concerning synonymy and nomenclature not addressed herein.

At one point during the early 1980s, revision of the Carolinas *Manual* was being considered (Radford, pers. comm.). Many of these corrections and suggestions were provided at that time in an effort to expedite publication of a second edition. Because of discovery of many new state and county records as well as many nomenclatural changes for Carolina plants since 1968, much of the original text now would have to be completely rewritten. Work toward that end is now underway by Alan Weakley, Peter White, and others, but a full-scale revision of RAB is not likely within the next decade (Weakley, pers. comm.). This publication does not account for additions to the Carolinas flora that have been discovered since 1968, nor was it written to fill this need. It was motivated by the hope that persons who use the Carolinas manual will find it to be even more helpful after these annotations are incorporated. Inclusion of as many of these annotations as possible in any available copy of the *Manual* should improve its usefulness and reliability until the time that a better means of identifying southeastern plant species becomes available.

Annotations and Identification Notes

Page	
vi	In treating format, haploid chromosome number is understood to be symbolized by the lower case letter "n" as used in the text descriptions. Entries used in illustrating manual format also indicate typeset, capitalization, and punctuation as well as the sequence of information provided in descriptions. This should have been stated concerning format, or else the parenthetical word "haploid" should have been "(Haploid ...)" and the word "Southeastern" in brackets should have been "southeastern."
vii	In last paragraph, "pentapolid" should be "pentaploid."
xiii	In the Contents, Glossary is listed for page "xlvii." The correct page number is "xlv."
xvi	In Pteridophyta key, line 5, delete the word "linear." Linear sori also occur in *Athyrium* (Aspidiaceae), and sorus position is more important than the shape, which is actually misleading, at this point.
xvi	In Pteridophyta key, line 7, "beween" should be "between."
xviii	Specimens of *Cercis canadensis* in flower but lacking leaves lead to Key 2. Fabaceae is not in Key 2; it is found in Key 6 for *Cercis* when leaves are present.
xviii	Specimens of *Cornus florida* in flower but lacking leaves lead to Key 2. Cornaceae is not in Key 2; it is found only in Key 4 (opposite leaves) and Key 6 (alternate leaves).
xviii	Specimens of *Passiflora incarnata* with leaves dissected more than 1/3 width of the blade lead to Key 8. Passifloraceae is not in Key 8; it is found only in Key 9.
xviii	Scapose specimens of *Salvia lyrata*, p. 911, lead to Key 10. Lamiaceae and other families having basically opposite leaves but some representatives that are scapose cannot be correctly keyed because these families are not included in Key 10.
xix	To the second half of couplet separating Rosaceae from remainder of Key 4 about three-fourths way down page, add a comma after "capsule."

xix Fruiting specimens of *Acer rubrum*, p. 688, lead to Aceraceae in Key 4 only if one assumes certain leaf characteristics are present, but *A. rubrum* (in AL) often produces fruits before the leaves.

xx The first half of couplet about one-third way down page "Petals united" vs. "Petals separate or absent" is not indented enough.

xx Specimens of *Decumaria barbara*, p. 520, with "leaves entire" do not lead to Saxifragaceae under this choice in Key 4. Saxifragaceae is found only under "Leaves toothed, lobed, dissected or compound" in Key 4.

xx Specimens of *Calycanthus*, p. 476, with "leaves entire" lead to Calycanthaceae in Key 4 only under the choice "Ovary inferior." Pistils are superior in *Calycanthus*, and Key 4 uses this feature under "Leaves toothed, lobed, dissected or compound" to lead to Calycanthaceae.

xxi In Key 5 Anacardiaceae can be reached only under "Leaflets or divisions toothed" for ternate-leaved specimens. Description of *Rhus radicans* on p. 676 states "Leaflets 3 ... entire to remotely serrate or shallowly lobed."

xxi Specimens of *Xanthorhiza simplicissima*, p. 453, with once-ternately or pinnately compound leaves do not lead to Ranunculaceae in Key 5. Ranunculaceae is found only under "Leaves 2–3x pinnately or ternately compound" in Key 5.

xxi Specimens of *Ptelea trifoliata*, p. 653, with "leaflets entire" lead to Fabaceae in Key 5. Rutaceae is found under "Leaflets or divisions toothed."

xxi Shrubby specimens of *Desmodium* or *Lespedeza*, pp. 604 and 613, do not lead to Fabaceae in Key 5. To the 10th line from the bottom of the page add "or a 1-seeded legume or loment" after the lead "fruit 2-or more-seeded."

xxi Specimens of *Gleditsia triacanthos*, p. 578, with "crenulate ... leaflets" and flowers actinomorphic do not agree with 2/3 of the necessary criteria for Fabaceae in Key 5.

xxii On lower page in the choice beginning "Stems not thick and fleshy," add a semicolon at end of first line to precede "nodes" on second line.

xxiv Specimens of *Vaccinium stamineum*, p. 814, which have corolla lobes shorter than the tube but rotate to campanulate shape, lead to Ericaceae only under "Corolla urceolate" or "Leaves evergreen" in Key 6. Add "campanulate or rotate" after "urceolate" in the first lead and delete "not urceolate" from the other half of this couplet.

xxiv Specimens of *Prunus serotina*, p. 565, a species with flowers perigynous and ovaries superior, do not lead to Rosaceae in Key 6 except under the choice "Ovary inferior" near the top of the page.

xxiv On lower page couplet "Corolla absent" vs. "Corolla present" is not properly indented. The latter is indented too far and is aligned with the couplet concerning "Stamens" which is itself slightly misaligned.

xxv Wording of line 4, "Stamens more than 10, monadelphous," does not correlate with that in line 5. Change line 4 to: "Stamens monadelphous; more than 10."

xxvi Fruits of *Alnus crispa*, p. 366, are "winged," but Betulaceae is not available under this lead in Key 6.

xxvi Fruiting specimens of *Halesia*, p. 827, lead to Styracaceae in Key 6 only if they are considered "not winged" or "dehiscent ..., an aggregate ..., or more than 1-seeded." None of these criteria are true of *Halesia* except the number of seeds per fruit, and this can be checked only with difficulty.

xxviii Specimens of *Chimaphila maculata*, p. 794, do not lead to Ericaceae either in Key 7 (if considered herbs with whorled leaves) or in Key 4 (if considered woody with whorled leaves). The only route to Ericaceae for this "subshrub" is via Key 10, and this will work only if the plant is considered an herb with "evergreen" leaves.

xxviii Specimens of *Cardamine diphylla* or *C. concatenata*, p. 505, which have opposite or whorled stem leaves, do not lead to Brassicaceae in Key 7. This family is not included in Key 7; such specimens can be keyed to Brassicaceae in Key 8.

xxix Both choices in couplet about two-thirds way down page leading to Acanthaceae and Scrophulariaceae state "flowers not in racemes, or spikes." Despite differences in punctuation of the pair of choices, this information is confusing and provides little help in making a choice at this point. The comma should be deleted after "racemes" in the first choice, and the "not" removed from the second choice, or else the whole phrase deleted from both choices.

xxix Choices in couplet leading to Verbenaceae and Lamiaceae about two-thirds way down page are poorly worded and confusing. Change to "Ovary smooth or ridged; style apical" vs. "Ovary distinctly lobed; style gynobasic."

xxx Specimens of *Houstonia* bear 2-locular fruits with 4–20 seeds per locule, but the lead to Rubiaceae near mid-page where such plants should key out states "fruit 2-seeded." Either change the word "-seeded" to "-locular," or delete reference to seed number.

xxx Specimens of *Rhexia*, p. 743, lead to Melostomataceae in Key 7 only if considered to have entire leaves. Leaves commonly are serrate or ciliate-serrate in *Rhexia*.

xxx Specimens of *Pinckneya pubens*, p. 978, do not lead to Rubiaceae in Key 7. Flowers are in cymes and sepals are 5 in *Pinckneya*, but Rubiaceae is not available under these choices.

xxx Specimens of Loganiaceae, p. 832, are characterized by superior ovaries; choice leading to this family in Key 7 under "Ovary inferior" near mid-page can be deleted so that "Leaves sessile" leads directly to Caryophyllaceae.

xxxi Specimens of *Alternanthera philoxeroides*, p. 422, are described as occurring in "Aquatic or very wet habitats," but to reach Amaranthaceae in Key 7 the choice "Plant terrestrial ..." must be followed. Change this choice, line 4, to "Plant terrestrial, or if aquatic or rooting at the nodes, the inflorescence peduncled or head-like."

xxxi Specimens of *Iresine*, p. 427, lead to Amaranthaceae in Key 7 under the choice including "fruits woolly and flowers imperfect." Fruits of *Iresine* are not woolly; the pubescent calyx is described as "white-sericeous."

xxxii Specimens of several *Sabatia* spp., pp. 836–838, bear flowers in diffuse terminal cymes or mixed panicles. Such plants lead to Gentianaceae in Key 7 only via the lead "Flowers not in loose open cymes ..."

xxxvii Specimens of *Lythrum*, p. 739, with alternate leaves and actinomorphic flowers do not lead to Lythraceae in Key 10 because the family is listed only under "Corolla strongly zygomorphic."

xxxvii Specimens of *Ludwigia uruguayensis* with 5 sepals, 5 petals, and stamens usually in whorls of 5 do not lead to Onagraceae in Key 10. This family is not found in this section of Key 10.

xxxviii Near mid-page "thryse" should by "thyrse."

xxxviii Specimens of certain *Saxifraga* spp., p. 528, should lead to Saxifragaceae at mid-page but do not. Add "or unlobed" after "Leaves lobed" in that choice.

xxxix Specimens of *Specularia*, p. 1002, do not lead to Campanulaceae in Key 10 except under the choice "Flowers in terminal racemes or spikes." The flowers are treated as "axillary" in the Campanulaceae key.

xxxix In couplet leading to Diapensiaceae, both choices are incorrectly punctuated after "terminal." A semicolon should follow the word "terminal" in the first choice and a comma in the second rather than *vice versa*.

xxxix Specimens of *Amsonia*, p. 845, do not lead to Apocynaceae at mid-page because Key 10 states "flowers in cymes with blue petals." Change this phrase to "flowers in panicles and petals blue."

xxxix Specimens of *Brasenia*, p. 452, with purple petals, do not lead to Cabombaceae in Key 10 because the only choice for this family requires "Petals white."

xl In couplet separating Droseraceae from other families above mid-page, "glanduar" in the second choice should be "glandular,"

xl Specimens of *Cnidoscolus stimulosus*, p. 661, with leaves lobed or cleft less than 1/3 the width of the blade do not lead to Euphorbiaceae in Key 10.

xli Schizocarp fruits of Boraginaceae, p. 878, do not lead one to that family in Key 11.

xli Capsular fruits of *Hydrophyllum*, p. 875, do not lead one to Hydrophyllaceae in Key 11.

xlii In line 12, "plan" should be "plant."

xlv In definition of Acaulescent, add "aerial" before "stem."

xlv In definition of Achene, add commas after "small" and "hard."

xlv Definition of Adherent is incorrect. Change to: Having unlike parts stuck together and appearing to be fused, usually due to surface wax or pubescence, but not truly adnate.

xlv Alveolate, a term used to describe seeds of *Lysimachia* on p. 821, should be defined: Honeycombed; having angular depressions separated by thin partitions.

xlv Amplexicaul, a term used to describe leaves of *Asclepias humistrata* on p. 852, should be defined: Clasping the stem.

xlv Anatropous, a term used in reference to ovules of *Luzula* on p. 280, should be defined: Said of an ovule whose opening (micropyle) is close to the point of stalk (funiculus) attachment.

xlv Androecium, a term used in many descriptions (e.g., *Schisandra*, p. 476), should be defined: All stamens of a flower taken collectively.

xlv Androgynophore, a term used in description of Passifloraceae on p. 733, should be defined: A stalk that bears both stamens and carpels and separates them from the receptacle.

xlv In definition of Androphore, insert "unfused portions of" before "stamens."

xlv In definition of Apiculate, "apicula" should be "apiculum" to agree in number with "a small, distinct point."

xlvi Apomict, apomictic, and apomixis, terms used in several descriptions (e.g., *Crataegus*, p. 558), should be defined: Reproducing without union of gametes; a type of asexual reproduction often involving the sexual structures.

xlvi Arachnoid; arachneous, terms used throughout text for pubescence, should be defined: Cobwebby by soft and slender entangled hairs.

xlvi Areolate; areole, terms used in reference to surfaces, should be defined: Divided into angular or squarish spaces; a small space (or area) delimited by veins in a reticulate pattern.

xlvi Auriculate, a term used in many descriptions, should be defined: Bearing prominent rounded basal lobes.

xlvi Basifixed, a term used in reference to *Heteranthera* on p. 272, should be defined: Attached or fixed by its base, usually applied to an anther attached by its base to the filament.

xlvi To definition of Beak, add "in Asteraceae, a slender apical elongation of the fruit body that elevates the pappus."

xlvi In definition of Bristle, delete "but" and add a comma in its place.

xlvi Bullate, a term used in the *Liatris* key, p. 1049, should be defined: Appearing blistered or puckered.

xlvii Cambium, a term used in the Angiospermae key, p. xvi, should be defined: A lateral meristem in vascular plants that produces secondary vascular tissues (xylem and phloem); annual increments of secondary xylem form the growth rings seen in wood of trees and shrubs from temperate climates.

xlvii Caruncle, a term used in description of *Luzula*, p. 280, should be defined: An outgrowth or appendage of a seed coat at or near the hilum.

xlvii In definition of Caulescent, add "leafy" before "stem."

xlvii Chalaza; chalazal, terms used in description of *Luzula*, p. 280, and elsewhere should be defined: The end of an ovule farthest from the micropyle; usually the basal end.

xlvii In definition of Cincinnus, add "of sessile flowers" after "cyme," delete "the," and add in its place "certain species of" before "Boraginaceae."

xlvii Clammy, a term used in Solanaceae key, p. 925, should be defined: Viscid, sticky, or gummy.

xlviii Coherent, a term commonly used in contrast with adherent, should be defined: Having like parts stuck together and appearing fused, usually due to surface wax or pubescence, but not truly connate.

xlviii Comose, a term used in Onagraceae key on p. 744 and Asclepiadaceae description on p. 848 and elsewhere, should be defined: Bearing a tuft of trichomes, usually at the apex.

xlviii Conduplicate, a term used in description of *Aristida spiciformis*, p. 98, should be defined: Folded longitudinally along the central axis so that the two halves of the ventral (upper) surface face each other.

xlviii Connivent, a term used for anthers of Solanaceae, p. 925, should be defined: Converging to the point of contact but without connation or cohesion.

xlviii Costate, a term used in description of Eriocaulaceae, p. 265, should be defined: Having ribs or riblike ridges.

xlviii Cotyledon, a term used in the Angiospermae key, p. xvi, should be defined: An embryonic leaf that functions in food storage or transfer within a seed and in certain cases becomes photosynthetic after the seed germinates.

xlviii Creeping, a term used with many vines and vinelike plants, should be defined: Running along at or near the ground surface and rooting; repent.

xlix Cusp, a term used in *Polygala* key on p. 656, should be defined: A sharp tooth or point, often rigid.

xlix To the definition of Cyathium add: "a cyathium contains several helicoid cymes of male flowers and one pedicellate female flower within its cupule on the rim of which one or more nectaries are usually present.

xlix Dictyostelic, a term used in the description of Aspidiaceae, p. 20, should be defined: A tubular network of amphiphloic vascular tissue with bundles wholly encased by endodermis and the spaces between them consisting of parenchymatous leaf gaps.

l E-, as used in "eciliate," p. 824, should be defined: Prefix denoting lack of some structure or part.

l Ectocarp, a term used in description of Araliaceae, p. 758, should be defined: Outermost layer of pericarp, often the mere skin of the fruit.

l Epigeous, a term used to describe gametophytes of *Lycopodium*, p. 5, should be defined: Growing or developing above ground; a type of seed germination in which cotyledons emerge from the seed coat and become elevated above the soil surface.

l Fistulose, a term used in description of *Swertia*, p. 842, should be defined: Hollow and cylindrical, as the leaf and stem of an onion.

l Flabelliform, a term used in description of *Lycopodium flabelliforme*, p. 7, should be described: Fan-shaped.

li Floccose, a term used in *Platanus* description, p. 531, and Asteraceae Key 9, p. 1012, should be defined: Loose tufts of long trichomes often persistent in leaf axils or on surfaces when other portions of the surface are glabrescent.

li Fruticose, used to describe suffrutescent in the Glossary, p. lix, should itself be defined: Shrubby or shrublike.

li Gemmae, a term used in descriptions of *Lycopodium* spp., p. 5, should be defined: Specialized asexual propagules capable of forming whole plants; in lower vascular plants, unlike the bryophytes, gemmae are associated with sporophytes rather than gametophytes.

li Glutinous, a term used in *Scirpus* key, p. 195, and description of *Carduus*, p. 1041, should be defined: Sticky; gluey.

li Gynobasic should be defined: Arising from the receptacle at or near the base of the carpels.

li Gynoecium, a term used in many descriptions including Apocynaceae on p. 845, should be defined: The carpels of a flower when referred to collectively, whether connate or not.

li Gynostegium, a term used extensively in Asclepiadaceae, pp. 847–858, should be defined: The structure formed by adnation of stamens to the gynoecium, as in *Asclepias*.

li Definition of Helicoid is incorrect and evidently confused with Scorpioid (which see). Change to: "An incomplete cyme with the sympodial axis coiled in one plane, the axis thus circinately coiled like a watchspring, as in the Boraginaceae."

li Hoary, a term used in definition of canescent (and there perhaps redundant) and elsewhere concerning pubescence, should be defined: Appearing like frost, grayish or grayish-white.

li Hypogynous, a term used in *Hypericum* key, p. 709, and in many descriptions, should be defined: Inserted upon the receptacle below the level of the carpels; thus, also descriptive of a flower with a superior ovary and no hypanthium.

li Imperfect flower on p. lii is out of alphabetical order; it should appear after Imbricate.

lii Infra-, applied in a number of instances in keys and descriptions, should be defined: A prefix indicating intermediate in position; in between.

lii Isobilateral, used in description of *Dicentra*, p. 482, should be defined: Pattern of symmetry in flowers and other parts of plants that can be divided into mirror-image halves by two perpendicular planes through the central axis; this pattern is considered to be actinomorphic.

lii Lamella, a term used in description of *Aplectrum*, p. 353, should be defined: A thin plate or layer; the septum between two cells of a compound ovary.

lii Lamina, a term used in description of Orchidaceae, p. 331, and elsewhere, should be defined: The blade or expanded portion, particularly in reference to leaves.

lii Lanose, a term used in Apiaceae Key 1, p. 761, and in other keys and descriptions, should either be defined or changed to "lanate:" Woolly, with long, intertwined, curly hairs.

lii Lanuginose, a term used in description of *Ponthieva*, p. 345, should be defined: Short woolly or cottony; downy, the hairs somewhat shorter than in lanate.

lii Laticiferous, a term used in description of *Euphorbia*, p. 668, and elsewhere (p. 847), should be defined: Bearing glands or ducts that produce milky sap, or latex.

lii Lenticel, a key term for *Vitis*, p. 695, should be defined: Corky areas on young bark, usually forming beneath the locations of stomata in epidermis as outer primary tissues become stretched by secondary growth.

lii Ligneous, a term used in Brassicaceae Key 1, p. 487, and in other descriptions and keys, should be defined: Woody.

liii Loment, a term used extensively in the Fabaceae beginning on p. 606 should be defined: A specialized legume with transverse separation of indehiscent, 1-seeded segments that commonly bear uncinate pubescence.

liii Micropyle (as micropylar), a term used in description of *Luzula*, p. 280, should be defined: The opening or pore (foramen) in the outer coats (integuments) of an ovule through which the pollen tube enters.

liii Mucronulate, a term used in descriptions of *Vaccinium* spp., p. 814, should be defined: Diminutive of mucronate.

liii Mycorrhizal (based on mycorrhiza), a term used in description of Ophioglossaceae, p. 10, should be defined: A symbiotic association between fungal hyphae and plant roots or other subterranean organs.

liv In definition of Ovule, add a comma after "fertilization."

liv Pandurate should be defined: Fiddle-shaped, the smaller end toward the point of attachment in the basic shape, thus resembling an obovate shape with a marked sinus along each basal side. This shape is very often treated as "obpandurate" due to the mistaken idea that the basal end must be wider than the apical end in terms used for plane figures.

lv Paraphyses (plural of paraphysis), a term used in description of *Polystichum*, p. 33, should be defined: Sterile filaments associated or intermingled with reproductive structures.

lv Pellucid, a term used in *Hypericum* key, p. 710, should be defined: Clear or transparent, usually applied to laminar surfaces in which internal oil glands allow light to pass through and appear as small dots (e.g., pellucid-punctate).

lv Penni-parallel, a term used in Monocotyledoneae key, p. xvii, should be defined: Venation pinnate with the lateral veins parallel to each other once diverged from the midrib.

lv In definition of Peri-, "suffix" should be "prefix."

lv Phyllodium (as phyllodia, phyllodial leaf, or phyllode), a term used in descriptions of *Sagittaria* spp., p. 52, and several genera of Apiaceae, pp. 783–787, should be defined: A somewhat expanded petiole that has the same form and function as a leaf blade.

lvi Prophyllum, a term used in description of Poaceae, p. 55, should be defined: A bracteole located at the base of an individual flower (e.g., the palea); a reduced leaf subtending a branch originating from a rhizome.

lvi Protuberance, a term used in describing the calyx of *Scutellaria* on p. 896, should be defined: A part or projection that bulges or swells out from a surface.

lvi Pseudocarp, a term used in description of *Calycanthus*, p. 476, should be defined: A fruiting structure derived principally from flower parts other than the ovary, often applied to an enlarged hypanthium or calyx tube that contains several to many unfused pistils.

lvi Pubescence is not in alphabetical order; it should appear after Puberulent.

lvi In definition of Pyxis, "circumscissle" should be "circumscissile." This error is recurrent throughout text.

lvi Raphids, a term used in description of Araceae, p. 257, should be defined: Needle-shaped crystals found in many plant cells, often in clusters or fascicles.

lvi In definition of Ray, "ligulate" is redundant for "strap-shaped;" the term is treated as "liguliform" on p. liii of Glossary, but ligulate should also be defined and included.

lvi Rhizome is not in alphabetical order; it should appear after Rhizomatous on p. lvii.

lvii Rostellum, a term used in description of Orchidaceae, p. 331, should be defined: A sterile modified stigmatic lobe in some orchids; a small beak.

lvii Rugulose, a term used in many descriptions and keys in Poaceae and Cyperaceae, pp. 55–255, should be defined: Finely rugose.

lvii Scaberulous, a term used in *Setaria* key, p. 126, and in descriptions of many other species, should be defined: Finely scabrous.

lvii Scandent, a term used in description of Fumariaceae, p. 482, should be defined: Climbing by stems, roots, or tendrils.

lvii Definition of Scorpioid is incorrect and evidently confused with Helicoid (which see). Change definition to: "An incomplete cyme with the sympodial axis coiled in an ascending spiral, the axis thus resembling a bedspring coil" (Lawrence 1951). The basic inflorescence in Boraginaceae is a helicoid (not scorpioid) cyme, but many authors persist in treating this form as scorpioid and generate unnecessary confusion by doing so.

lvii In definition of Segment, insert "or the unit portion of the perianth, corolla, or calyx" after "compound" and before "any."

lviii Setose is not in alphabetical order; it should follow Setiform.

lviii In definition of Silicle, "as long as broad" should be "as broad as long," and "never more than twice as long as broad" should be added.

lix Stipel, a term used in descriptions of many Fabaceae, pp. 569–645, should be defined: A basal appendage of a leaflet analogous to a stipule.

lix In definition of Stipule, add "One of" at the beginning to cause agreement in number.

lix Stomium, a term used in description of Pteridaceae, p. 15, should be defined: A pore or opening; the lip-cell region of a fern sporangium through which dehiscence occurs.

lix Strobilus (plural = strobili) should be defined: A cone; a shortened stem axis bearing an aggregation of sporophylls, sporangiophores, or ovuliferous scales.

lix Sub-, used as a prefix before many basic terms (e.g., subcylindric, p. 869) should be defined: A prefix indicating approximately, almost, or nearly.

lix In the definition of Suffrutescent, change the word "Diminutively" to "Slightly or obscurely," and change period to comma and add "subshrubby" at the end.

lix Suffruticose, a term used in *Hypericum* key, p. 709, and Solanaceae key, p. 925, should be defined: Very low and woody; diminutively shrubby.

lix In definition of Teeth, either change word to "Tooth" or make plurals of "protuberance," "serration," and "dentation" to cause agreement in number.

lix In definition of Ternate, use of apostrophe in "three's" is incorrect. Change to "threes" or use numeral (e.g., "3's").

lx Definition of Triquetrous is incorrect. Change to: "Three angled in cross section, as the stems of Cyperaceae, with the surfaces often concave."

lx Ul-, as an internal syllable placed between a root word and the suffixes "-ate, -ose, and -ous" indicates a diminutive condition: e.g., crenulate, lanulose, scaberulous, serrulate, uncinulate, etc.

lx Uncinate, a term used for *Desmodium* fruits on pp. 604–613 and elsewhere, should be defined: Erect trichomes or spines with short hooked apices, or pubescence made up of such trichomes.

lx Uncinulate, a term used for description of fruit surfaces in *Gaura* on p. 754 and *Desmodium* on p. 604, should be defined: Puberulent with diminutive uncinate trichomes.

lx Usually, a nontechnical word listed (but not actually defined) in the Glossary, should be deleted.

lxi Vestiture, a term used in *Physalis* key, p. 927, and elsewhere, should be defined: Anything on or arising from a surface that causes it to be other than glabrous; papillae, scales, blisters, warts, etc. as well as trichomes serve as plant vestiture; also known as vesture.

6 In description of *Lycopodium appressum*, line 6, "eaves" should be "leaves;" line 7, "Strobili ... 5–10 cm broad" should be "5–10 mm broad."

7 *Lycopodium flabelliforme* (Fernald) Blanchard = *L. digitatum* Dillen. ex A. Braun.

10 In *Botrychium* key, line 3, "pinae" should be "pinnae."

13 In description of *Osmunda cinnamomea*, line 2, "2pinnatifid" should be "2-pinnatifid."

15 In description of *Trichomanes boscianum*, line 4, "pinae" should be "pinnae."

17 In description of *Dennstaedtia punctilobula*, line 3, "4–10 cm" should be "4–10 dm."

21 In description of *Woodsia ilvensis*, line 2, "concolorus" should be "concolorous."

27 In *Thelypteris* key, the couplet "rachis with many dark scales beneath" vs. "rachis with only a few pale scales beneath" leading to *T. phegopteris* and *T. hexagonoptera*, respectively, should be changed to "rachis scales numerous and dark" vs. "rachis scales few and pale."

34 In key to *Pinus* species and various descriptions on pp. 36 and 38, reference to leaves or foliage in "2s, 3s, or 4s" is incorrect. Plurals of numerals require an apostrophe; change to "2's, 3's, or 4's" as appropriate; see also Glossary annotation for Ternate, p. lix.

36 Bark of branches of *Pinus glabra*, p. 38, varies from smooth to ridged depending on age and size, but one lead in the *Pinus* key allows only "smooth" for this species.

36 In description of *Pinus serotina*, line 3, "fema le" should be "female."

41 In description of Cupressaceae, line 1, "pairs of" should be "pairs or."

42 The illustration of *Juniperus virginiana* shows a cedar-apple rust fungus telial gall and a bagworm moth larval case. These parasite-caused structures are not mentioned in the description and are not features of the plant alone; thus, they can confuse and mislead.

51 In key choice leading to *Sagittaria graminea*, "cresent-shaped" should be "crescent-shaped."

51 Key choice leading to *Sagittaria teres* states "emersed leaves ... less than 2 mm wide." The description, p. 53, states "Emersed leaves ... to 0.5 cm wide."

52 *Sagittaria subulata* Buch = *S. subulata* (L.) Buch.

54 In description of *Sagittaria engelmanniana*, line 1, "tyyically" should be "typically;" line 5, "prmoinent" should be "prominent."

54 Key choice leading to *Egeria* states "principal leaves 1.2–4 cm long." Description states "Leaves ... 1.2–2.5 cm long."

54 Key choice leading to *Elodea* states "principal leaves 0.5–1.7 cm long." Descriptions of *E. canadensis* and *E. nuttallii*, p. 55, state "Leaves 8–15 mm long" and "Leaves 3–15 mm long," respectively.

55 In description of *Elodea canadensis*, line 3, "through" should be "throughout."

57 In Poaceae Key 2, p. 57, near mid-page, change the couplet "Plant dioecious" and "Plant not dioecious" to "Plants..." A single plant can be unisexual and part of a dioecious species, but an individual can hardly be dioecious.

57 Specimens of *Uniola*, p. 65, without sterile lemmas below the fertile (not uncommon in AL specimens) do not lead to *Uniola* in Poaceae Key 2. Woodland species of this genus are now treated in *Chasmanthium* (Yates 1977); see pp. 65-66.

58 Glumes of *Danthonia sericea* are sometimes 7-nerved, and this leads to *Avena* in Poaceae Key 4.

58 Key choice leading to *Polypogon* in Poaceae Key 5 states "Glumes ... 1.5–2.5 mm long." Description on p. 102 states "Glumes ... 1.5–2 mm long, awns 3–8 mm long."

58 Key choice leading to *Lagurus* in Poaceae Key 5 states "Glumes plumose, 10 mm long." Description on p. 102 states "Glumes ... pilose ..., body 2–4 mm long, awn plumose, 3–7 mm long."

58 Key choice leading to *Sporobolus* in Poaceae Key 5 states "Glumes awnless." Description of *S. vaginiflorus* on p. 105, which should key under that choice in Key 5, states "Glumes ... acute to short awned."

58 Key choice "Callus beard absent" on last line can lead to *Brachyelytrum* on p. 59. Description of *B. erectrum* on p. 115 states "callus short bearded." Such entries in the key seem to have been made to facilitate correct identification when certain key features are hard to observe; this can be confusing when a plant specimen is compared with such a description while the key choices that led to that description are still in mind.

59 Choice in Key 6 leading to *Bouteloua* states "spikes usually less than 10 cm long." Description on p. 120 states "spikes ... 1–3 dm long."

67 Key choice leading to *Eragrostis ciliaris* states "Lemma 1–1.3 mm long." Description on p. 68 states "lemmas ... 1.1–1.3 mm long."

65 Lemma lengths given in the primary couplet in *Uniola* key refer to fertile lemmas only. Add "Fertile" to beginning of each choice in this couplet.

65 *Uniola laxa* (L.) BSP. = *Chasmanthium laxum* (L.) H.O. Yates.

66 *Uniola sessiliflora* Poiret = *Chasmanthium sessiliflorum* (Poiret) H.O. Yates; *U. latifolia* Michaux = *Chasmanthium latifolium* (Michaux) H.O. Yates; *U. nitida* Baldwin = *Chasmanthium nitidum* (Baldwin) H.O. Yates.

71 Grains of *Eragrostis spectabilis* are not described though important in the key to species. Add: Grain dark brown, ovoid, 0.6 mm long (Hitchcock 1950).

72 Key choice leading to *Bromus ciliatus* states "lemma awn less than 2.5 mm long." Description on p. 74 states "awns 1.5–5 mm long."

82 Some overlap exists in primary key characteristic leading to *Festuca octoflora* and *F. sciurea*, respectively. Key indicates more or less than 2.5 mm for first glume length, but descriptions indicate as short as 2 mm for the former species and as long as 2.6 mm for the latter. *Festuca* spp. 1-3, pp. 82–83, are usually treated under synonyms in *Vulpia* that are listed at the ends of descriptions.

82 In description of *Festuca octoflora*, line 8, add period and comma after "Ala."

83 In description of *Festuca sciurea*, line 2, "10–15 mm lonff" should be "10–15 mm long."

85 Key choice leading to *Lolium temulentum* states "Glume as long as spikelet." Description states "Spikelets ... 2–2.5 mm long" and "First glume ... 10 mm long, 2nd glume 12–13 mm long; lemma awns 10–14 mm long." It is not clear what should be measured to make the comparisons required to use the key; lemma body length is not given.

85 Plants of *Lolium multiflorum* are only weakly perennial; sometimes biennial or even annual.

86 In *Agropyron* key, information in the couplet "lemma awned" vs. "lemma awnless" leading to *A. repens* and *A. trachycaulum*, respectively, is not reflected in the descriptions. The former species description states "lemma ... 1–11 mm long," and the latter states "awns 1.5–6 mm long or absent."

98 Key choice leading to *Aristida virgata* states that "Lateral awns or shortest awn more than 7 mm long." Description on p. 101 states "awns ... 5–20 mm."

101 In description of *Aristida longispica*, line 1, "clums" should be "culms."

110 In *Agrostis* key, remove period and add dotted line in choice leading to *A. perennans*.

119 Key choice leading to *Spartina cynosuroides* states "spikelets more than 9 mm long." Species description on p. 120 indicates "8.5 mm" as minimum spikelet length.

122 Key choice for *Leersia oryzoides* states "Spikelets 4–5.5 mm long." Description on p. 123 states "Spikelets 4.2–5 mm long."

133 Key choice for *Paspalum vaginatum* states "Spikelets glabrous, largest 3–4 mm long." Description on p. 134 states "Spikelets ... 2.5–3 mm long."

133 Key choice for *Paspalum dilatatum* states "Racemes 3–9." Description on p. 134 states "Racemes 3–7."

138 In description of *Digitaria ischaemum* var. *violescens*, line 1, "Spikleets" should be "Spikelets."

141 Key choice leading to *Sacciolepis striata* states "spikelets 3.5–5 mm long." Description states "Spikelets 4–5 mm long."

142 In *Panicum* Key 1 the choice leading to *P. gymnocarpon* states "Fertile lemma 1/3 length of sterile lemma." Description on p. 148 states "2.2 mm" and "5 mm" as respective lengths for these structures. Key choice should be changed to "Fertile lemma less than 1/2 length of sterile lemma."

142 In *Panicum* Key 1 the second choice under "Panicle more than 2 cm broad" is not properly indented. It is indented too far and is aligned with couplet "Spikelets secund" vs. "Spikelets not secund."

169 Scale and nutlet lengths given for *Cyperus filicinus* and *C. polystachyos* in key do not cover the same limits as stated in the descriptions on p. 172.

170 *Cyperus* spp. 1–3 are commonly treated under synonyms in *Kyllinga* indicated at the ends of the species descriptions.

183 Choice leading to *Eleocharis rostellata* states "nutlet 2 mm or more long." Description on p. 186 states "Nutlets ... 1.5–1.8 mm long."

183 Choices leading to *Eleocharis vivipara* vs. *E. tuberculosa* and *E. tortilis* do not reflect nutlet lengths given in descriptions on p. 188 for these species. If descriptions are correct, the couplet should be changed to "Nutlet less than 0.8 mm long" vs. "Nutlet more than 1.2 mm long."

190 Key choice leading to *Bulbostylis capillaris* states "Spikelets ... rarely solitary." Description of this wide-ranging species states "Spikelets solitary." Change "rarely solitary" to "or solitary" in the key.

195 In couplet separating *Scirpus robustus* from rest of key, spikelet lengths do not agree with those given in descriptions for *S. robustus*, p. 197, "1.5–4.5 cm," nor for *S. lineatus*, p. 198, and *S. divaricatus*, p. 201, of up to "12 mm" and "15 mm," respectively. Change key choices to "Spikelets 15 mm or more long" vs. "Spikelets usually less than 10 mm long."

201 In description of *Scirpus divaricatus*, line 8, "minutley" should be "minutely."

203 Couplet "Spikelet 1-fruited" vs. "Spikelet 2-fruited" leading to *Rhynchospora* species 13–17 oversimplifies the case for *R. glomerata*. Add "only" to first choice and change second choice to "Spikelet usually 2-fruited, occasionally 1-fruited."

203 Choices leading to *Rhynchospora capitellata*, p. 208, and, later in the key, to *R. filifolia* and *R. pleiantha* state "Nutlet ... obpyriform." The shape shown in the figures is "pyriform." Much confusion exists due to misuse of these two terms, but, in short, in pears and fiddles (pyriform and pandurate) the small end is the point of attachment of the basic shape; the prefix "ob-" denotes the inverted shape just as is done with the shapes when the big end happens to be toward the base in the basic figure. See annotation for Pandurate, p. liv.

205 In figure title of *Rhynchospora corniculata*, spacing of the letters "rn" in the epithet make them resemble an "m" on first inspection.

220 In *Carex* Key 6, last line, the "g" in *C. glaucescens* resembles an "a."

240 In description of *Carex conoidea*, line 6, "geeenish" should be "greenish."

267 Figure of male flower of *Eriocaulon decangulare* states "x 1/6." Change this to "x 6."

266 Scapes of *Eriocaulon decangulare* (from FL) sometimes are more than 9 dm long, and leaf sheaths more than 15 cm.

268 In the figure title of *Tillandsia usneoides*, the final "a" in *Tillandsia* resembles a "q."

268 Couplet leading to *Commelina virginica* and *C. erecta*, respectively, is inconsistent with descriptions on p. 269. Add "with reddish or yellowish bristles" to the first choice, and change the second to "leaf sheath ciliate with pale bristles."

268 Key choice leading to *Commelina diffusa* states "spathe 10–25 mm long." In the species description, p. 269, line 4, "1.7–2.5 mm" should be "1.0–2.5 cm."

271 In key choice leading to *Tradescantia subaspera*, add a comma after "opened."

282 Specimens of *Smilax ecirrhata*, p. 285, commonly have leaves appearing whorled ("subverticillate"), but this choice in the Liliaceae key leads to Key 1 which does not include *Smilax*.

282 In first choice in Liliaceae Key 5, "helicoid" should be "scorpioid." (See annotations for these terms, pp. li and lvii).

283 In the figure title for *Asparagus officinalis*, the "ff" in "officinalis" looks like "tf."

284 In the description of *Asparagus officinalis*, line 5, "thoughout" should be "throughout."

284 In *Smilax* key, leaf shapes "oblong or pandurate" vs. "lanceolate" given for choices leading to 9. *S. auriculata* and 10. *S. laurifolia*, respectively, are misleading. In the choice leading to *S. auriculata*, "pandurate" should be "hastate" or "obpandurate," and leaves of *S. laurifolia* available via the second choice are described on p. 287 as "oblong or oblong-lanceolate or oblong-linear."

284 In *Smilax* key, the choice beginning "Peduncles 1.5x or less as long as the petioles ..." needs a comma after "leaves."

286 In the figure title for *Smilax bona-nox*, the "b" in *bona-nox* looks like one struck-over with a "p."

289 In description of *Trillium*, the phrase "Flowers perfect, sessile or pedicellate" is inconsistent with the key which states "Flowers peduncled" vs. "Flowers sessile" and refers to "peduncle" in four key choices. The flower stalk in *Trillium* is considered homologous with the pedicels in *Smilax* and *Medeola*; thus, the term of preference here is "pedicel." This is contrary to a restrictive definition of the term peduncle as the support stalk of any solitary flower as treated by many authors. Many solitary flowers are borne on peduncles (e.g., *Duchesnea*, *Hibiscus*); some on pedicels (e.g., *Trillium*); and some on both peduncle and pedicel (e.g., *Zephyranthes*, acaulescent species of *Viola*). The location of bracts provides a means of discriminating among these various types; in *Trillium* the "bracts" are foliaceous and commonly are referred to as "leaves." Descriptions for *Trillium* species 5–10 on pp. 290–293 appropriately refer to the flower stalks as "Pedicels."

290 *Trillium cuneatum* Raf. var. *luteum* (Muhl.) Ahles = *T. luteum* (Muhl.) Harbison (Freeman 1975); *T. viride* Beck and Coastal Plain plants of *T. cuneatum* Raf. (both sensu Ahles) = *T. maculatum* Raf.; *T. discolor* Wray = *T. discolor* Wray ex Hooker; *T. lanceolatum* Boykin = *T. lancifolium* Raf.; *T. cernuum* L. (sensu Ahles) = *T. rugelii* Rendle; *T. erectum* L. var. *vaseyi* (Harbison) Ahles = *T. vaseyi* Harbison (Samejima and Samejima 1987).

309 Description of *Hemerocallis*, line 3, states "forked into 2 bracteate racemes." The inflorescence is a pair of scorpioid cymes, or false spikes.

309 Choices in *Lilium* key leading to *L. superbum* and *L. michauxii*, respectively, state "Leaves ... lanceolate to oblanceolate" vs. "Leaves ... spatulate." This does not agree with the descriptions of these species, p. 311, which state "Leaves ... lanceolate to elliptic" and "Leaves ... oblanceolate," respectively.

320 Description of *Zephyranthes*, line 4, states "Perianth ... 0.5–2.5 cm long." Perianth lengths indicated both in the key and in species descriptions are all longer than these limits.

320 Perianth of *Zephyranthes simpsonii* must be "deep pink" to agree with key; description states "usually pink." In some plants it is white with purple tips.

325 In description of *Sisyrinchium*, line 5, "8–15 cm" should be "8–15 mm."

327 Key choice leading to *Iris virginica* states "Bracts subtending spathes less than or slightly longer than the spathes, or absent." Description on pp. 328–329 states "Outer bracts subtending spathes slightly shorter to nearly 2x as long as spathes." Change key choice to "less than up to 2x as long as the spathes, or absent."

327 In *Gladiolus* key, last line, "2.5 mm long" should be "2.5 cm long."

327 *Gladiolus* species "1. *G.* x *gandovensis*" should be "*G.* x *gandavensis*," as correctly spelled in the key.

356 Pistillate inflorescences in families 51–56 should not be treated as "catkins." They are better treated as spikes even when catkin-like. Catkins consist of cymules of male flowers on elongate monopodial axes that fall as units after anthesis, and neither this structure nor this fate is true of pistillate structures in catkin-bearing taxa.

363 In *Carya* key the statement "husk sutures not winged or keeled" in the first couplet is misleading for *C. pallida* and *C. ovalis*.

363 Fruiting twigs of *Carya illinoensis* sometimes have leaves with 9 or fewer leaflets each. Such specimens lead to *C. myristicaeformis* in the *Carya* key.

365 In the description of *Carya tomentosa*, line 3, the broken numeral after "lanceolate" should be a "3."

367 In the description of *Corylus cornuta*, line 4, add a period after "W.Va."

373 In *Quercus* key, line 11, "Leaves with 10 of more ..." should be "Leaves with 10 or more ..."

377 In the figure title for "*Quercus muehlenbergii*" the first "e" is missing from "muhlenbergii." This epithet is variously spelled in the literature; most recent publications accept *Q. muehlenbergii* as correct.

385 In the description of *Quercus laurifolia*, line 9, "Mrach" should be "March."

392 Figure of *Urtica chamaedryoides* indicates whorled leaves, but key and description on p. 393 indicate only opposite leaves for *Urtica*.

398 Leaves of *Asarum*, p. 399, appear to be opposite; specimens key to Aristolochiaceae under this choice in Key 7 on p. xxx. Add "or opposite" after "alternate" to family description, line 1.

400 In the *Hexastylis* key, line 2, "saggitate" should be "sagittate."

400 In description of *Hexastylis arifolia*, line 2, "16–20 cm long" should be "6–20 cm long."

403 In *Rumex* key, "Verticels" should be "Verticils," and in line 14, "margins" should be transposed with "entire."

407 Key choice leading to *Polygonum setaceum* states "ocreal cilia usually 10 mm or more long." Description on p. 412 states "cilia 6–10 mm long."

407 Key choice leading to *Polygonum hydropiperoides* states "Leaves less than 15 mm wide." Description on p. 412 states "Leaves ... 0.5–2 cm wide."

409 Leaves of *Polygonum pensylvanicum* (in AL) are sometimes strigillose underneath.

409 Ocreae of *Polygonum pensylvanicum* (in AL) not uncommonly are short ciliate or strigillose at the margin; cilia are about 0.5 mm long.

412 In description of *Polygonum convolvulus*, line 2, "cordate-saggitate" should be "cordate-sagittate."

425 In description of *Amaranthus pumilus*, line 3, "indehsicent" should be "indehiscent."

427 Nyctaginaceae can be reached in Key 7 on p. xxx either under "Ovary inferior" or "Ovary superior," but only the latter condition is true of the family. The description should read "ovary superior, but appearing inferior due to adhesion of calyx tube to pericarp."

430 In Aizoaceae key, the choice leading to *Tetragonia* is "Leaves opposite." Description on p. 431 states "Leaves alternate." Leaves may be either opposite or alternate in this genus.

431 Decumbent stems of *Mollugo verticillata* sometimes root at the nodes (in AL).

432 In description of Portulacaceae, line 3, "circumscissle" should be "circumscissile."

433 Ditto, description of *Talinum*, line 4.

434 Ditto, description of *Portulaca*, line 3; also each of the *Portulaca* species except *P. coronata*.

435 Key choice "Sepals united; ovary stipitate or sessile" is indented too far and is misaligned with the couplet concerning "Styles ..."

447 Specimens of *Silene ovata*, p. 448, lead to the correct identity in the *Silene* key only under the choice "calyx pilose." The calyx is not pilose; it is minutely pubescent in this species.

459 Correct authority for *Viorna glaucophylla*, a synonym of *Clematis glaucophylla* Small, is "(Small) Small."

459 *Clematis dioscoreifolia* Levl. & Vaniot = *C. terniflora* DC.

462 Petals of *Ranunculus recurvatus*, p. 465, are sometimes slightly longer than the sepals in fresh specimens; they appear to shrink more on drying than sepals.

466 In description of *Ranunculus repens*, line 2, "2–10 dm wide" should be "2–10 cm wide."

471 In description of Menispermaceae, "*Calycocarpum lyoni*" should be "*Calycocarpum lyonii*."

473 In description of Magnoliaceae, "Hermaphroditic trees" is inappropriate; the term hermaphroditic is better applied (and limited) to flowers. (Likewise, monoecious trees are bisexual but should not be called hermaphroditic.) Change to "Trees with perfect (or hermaphroditic) flowers."

476 In description of *Asimina parviflora*, line 4, the fruit dimensions given are too small to accomodate the seed size. Fruits often are longer than "3 cm" and wider than "1.5 cm."

479 Change "*Lindera melissaefolium*" to "*L. melissifolia*." *Lindera* is feminine in gender.

480 Certain white-petalled specimens of *Argemone* (from AL and FL) that key to *A. albiflora* have yellow latex and thus do not agree with description. This group is in need of study.

486 In description of Brassicaceae, line 4, "stamens 6, 2 long and 4 short" should be "stamens 6, 4 long and 2 short." There are 4 long stamens in a tetradynamous androecium.

487 In Brassicaceae Key 1, in the last choice that leads to *Cardamine*, "one- or twice-palmately" should be "once- or twice-palmately."

489 In the couplet "Stems procumbent ..." vs. "Stems erect ..." in Brassicaceae Key 4, the word "procumbent" should be "repent," and "rooting at the nodes" should be deleted.

489 Last couplet in Brassicaceae Key 4 should be changed to "Upper stem leaves lobed or more deeply divided" vs. "Upper stem leaves not lobed nor deeply divided." Stem leaves of *Cardamine parviflora* are divided more than 1/3 width of the blade (i.e., more than "lobed") and can be reached only via the first choice.

499 In description of *Sisymbrium altissimum*, lines 4–5, "Silique ... 7–8 mm long" should be "7–8 cm long."

504 In description of *Barbarea vulgaris* var. *vulgaris*, "railraods" should be "railroads."

505 In *Cardamine* key, the choice under which *C. clematitis* is found states "Leaves, at least the basal ones, deeply lobed or dissected." Species description on p. 508 states "Basal leaves orbicular, cordate, crenate or remotely lobed."

516 Description of genus *Drosera* states "Annual or perennial ... herbs," but species descriptions, pp. 517–518, do not indicate duration or growth form.

522 Key choices leading to *Philadelphus inodorus* and *P. hirsutus*, respectively, state "Stigmas and styles separate" vs. "Stigmas and styles fused." Description of *P. inodorus*, lines 4 and 5, states "stigma and style branches 3–4 mm long, united style 5–6 mm long." Thus, it is not clear what precise structure the key refers to as being separate or fused. Both species have a common (or united) style; the key should be changed to: "Stigmas on style branches 3–4 mm long" vs. "Stigma unbranched, less than 3 mm long."

529 Leaves of *Saxifraga careyana* are finely strigose to glabrescent above.

532 In Rosaceae key, to the second half of the couplet that leads to *Alchemilla* add "or pedunculate" after "pedicellate."

532 Choice leading to *Rubus* at mid-page in Rosaceae key states "Stipules not adnate to petiole." Description on p. 537 states "stipules ... adnate to petioles." Stipules are narrow, inconspicuous, and persistent; the key more accurately describes fusion to petiole.

532 Fruiting specimens of *Sorbus arbutifolia*, p. 558, often appear to have simple rather than compound corymbs due to failure of some flowers to form fruit and subsequent loss of those branches. Such specimens do not lead to *Sorbus* in Rosaceae key I.

532 In Rosaceae key I. Fruiting Specimens, the first listing of "*Cratiaegus*" should be "*Crataegus*." (This error appeared only in the first few printings.)

533 In *Fragaria* key, line 7, "inbedded" should be "imbedded."

533 In description of *Duchesnea indica*, line 4, "pedicels" should be "peduncles."

535 Key choice leading to *Potentilla recta* states "Leaflets usually 7–9, rarely 5." Description states "leaflets 5–7, rarely 9." Change key choice to "Leaflets usually 7, rarely 5 or 9."

535 In description of *Potentilla canadensis*, line 7, and *P. simplex*, line 2, "pedicel" should be "peduncle."

541 Add asterisk(*) after *Rubus illecebrosus* Focke; the species is illustrated on p. 542.

543 Flowering key choice for *Geum virginianum* states "Petals as long as or shorter than the sepals." Description states "Petals ... shorter than the sepals."

548 In *Rosa* key, the primary feature separating *R. wichuraiana* and *R. multiflora* is leaflet number. Both descriptions on p. 550 state "Leaflets 7–9," and the description of the former species, reached only via the key choice "Leaflets usually 9," states "The common form ... "Dorothy Perkins" ... usually 7 leaflets." The key information is misleading, and single quotation marks (e.g., 'Dorothy Perkins') should be used for a cultivar.

556 Key choice leading to *Malus angustifolia* states "Leaves crenate." Description on p. 557 states "Leaves ... crenate to serrate," and illustration shows leaves more serrate than crenate. Add "usually" after "Leaves" in this key choice.

560 Key choice leading to *Crataegus crus-galli* states "Pyrenes mostly 2," but description on p. 562 states "pyrenes 3–5." Change description to "pyrenes 1-3."

564 In description of *Amelanchier arborea*, line 4, "petals 1–1.7 mm long" should be "petals 1–1.7 cm long."

565 In description of *Amelanchier canadensis*, lines 4 and 5, "Sepals ... 1.5–3 cm long" should be "Sepals ... 1.5–3 mm long."

565 Key choice leading to *Prunus umbellata* states "fruit usually blue," and the description on p. 566 states "Drupe black, very glaucous." The key can be improved, and the apparent ambiguity avoided, by changing the key choice to "fruit glaucous, appearing bluish."

569 In description of Fabaceae, line 7, the reference to "petals ... adhering along one margin" should be "petals ... cohering ..."

569 Placentation type described for Fabaceae seems to be slightly misleading. In unicarpellate pistils that bear 1-many ovules along the ventral suture, the placenta is often termed marginal rather than parietal. This enables emphasis to be placed upon the primitive character of such pistils when compared with the more advanced condition of syncarpous pistils with typical parietal placentae.

569 In sentence about *Parkinsonia aculeata*, "papillionaceous" should be "papilionaceous," and add a comma after "flowers" on same line.

570 In Fabaceae Key 2, line 19, the period after "Petal 1" should be a comma.

570 Specimens of *Petalostemum pinnatum*, p. 601, (from AL and FL) are commonly trifoliolate. This condition leads to Key 3 which does not include *Petalostemum*; this genus is included only in Key 8.

571 In Fabaceae Key 5, line 13, add a comma after "1-seeded."

571 In Fabaceae Key 5, choice leading to *Lespedeza* states "Petals not blue, lavender or violet." Description on p. 613 states "corolla ... violet, purplish, roseate." This wording causes many species of *Lespedeza* to key to *Psoralea*.

571 In Fabaceae Key 5, the choice leading to *Stylosanthes* states "Petals bright butter yellow." The other half of this couplet states "Petals pink ..., at least not bright butter-yellow." Such terms for color as "butter-yellow" should either be consistently hyphenated or not; the hyphenated form is preferable.

572 Key choice leading to *Vigna* in Key 6 falls under the statement "Standard 2 cm or less long." Description of *V. unguiculata* on p. 639 states "the standard 1.5–2.5 cm long." Since standard petals of *Centrosema* and *Clitoria* are 2.5 cm or more long, the first couplet in Key 6 should state "Standard 2.5 cm or more long" vs. "Standard 2.5 cm or less long."

572 Ditto concerning *Pueraria* in Key 6 and description on p. 642.

573 Key choice leading to *Galactia* in Key 8 falls under "leaflets 5–7." Description and key on p. 643 for *G. elliottii* state "Leaves pinnately 7–9 foliolate" and "Leaves 5–9 foliolate," respectively. Add "usually" to the statement in Key 8, and change the description to "Leaves pinnately 5–9 foliolate."

576 Add asterisk(*) after *Cassia hebecarpa* Fernald; the species is illustrated on p. 577.

580 Key choice in *Baptisia* key leading to *B. lanceolata* states "ovary glabrous or sparsely pubescent." Description on p. 581 states "ovary densely pubescent." This inconsistency together with relatively large flowers cause this species to fall under the contrasting choice which incorrectly leads to either *B. cinerea* and *B. bracteata*.

581 Specimens of *Baptisia lanceolata* (from AL and FL) bear legumes with body length up to 2.5 cm and diameter of 1.6 cm.

585 In *Crotalaria* key, "Leaftlets" should be "Leaflets" in line 4 under the choice "Leaves 3-foliolate."

588 In *Trifolium* key, the last line of the choice leading to *T. arvense* is not properly spaced vertically. There are 4 blank lines in this key, whereas most keys have continuous spacing vertically.

589 In description of *Trifolium lappaceum*, last line, "ver rare" should be "very rare."

597 In description of *Lotus*, line 2, "stipules minute, gland-like" is incorrect for *L. corniculatus*. Stipules are ovate and foliaceous, and the leaves 3-foliolate.

597 In *Amorpha* key, information about pubescence of leaves and young branches is omitted from the second half of the first couplet. Add "leaves and young branches usually glabrous or glabrate, rarely strigillose, short-pubescent or pilose."

602 Leaflets of *Coronilla varia* (in AL) often measure 8 mm or less in length.

604 Key choice leading to *Desmodium* Key 4, which includes *D. nuttallii*, states "Loment segments (1) 3–5 (6), the segments usually angular; corolla mostly 6–9 mm long." Description on p. 611 states "Loment ... with 2–4 segments, rounded both above and below" and "petals ... 4–7 mm long." These size ranges and qualifications agree more closely with the key data leading to Key 3, but *D. nuttallii* is not available in that key.

605 Primary couplet separating *Desmodium floridanum* from rest of Key 4 is poorly worded and contains size limits inconsistent with species descriptions. Change first half of couplet to "Stipules persistent, conspicuous, usually 6 mm or more long" and second half to "Stipules soon deciduous, usually 5 mm or less long."

619 Description of *Sesbania exaltata*, line 9, states "Legume ... 30–40 seeded." Fabaceae Key 7 on p. 573 states "seeds 20 or more" in the choice leading to *Sesbania*. If as few as 20 seeds occur in *Sesbania* legumes, the description should reflect this.

624 Key choice leading to *Tephrosia virginiana* and the description on p. 625 state "stamens monadelphous" for this species. In some specimens (from AL to GA) the upper stamen is free. The first choice should be changed to "stamens usually monadelphous ... (rarely diadelphous)."

636 In description of *Clitoria mariana*, lines 9 and 10, the "1–14." at end of line 9 should be "1–1.4" referring to the "cm" on the next line.

636 In *Rhynchosia* key, the first species "2. *R. reniformis*" should be "1. *R. reniformis.*"

643 "*Amphicarpa*" should be "*Amphicarpaea.*"

645 In descriptions of both *Linum usitatissimum*, line 1, and *L. striatum*, line 1, "3–10 cm" should be "3–10 dm."

653 In description of *Zanthoxylum clava-herculis*, line 2, delete the comma after "crenate."

661 Description of Euphorbiaceae states "In our flora only *Euphorbia* has milky sap, plants of our other genera have watery sap." Both *Cnidoscolus* (native) and *Sapium* (naturalized) also have milky sap.

661 In Euphorbiaceae key, the lead to *Croton* states "Fruit 3-locular, 3-seeded, dehiscent." In *C. monanthogynus*, p. 663, the fruit is "1-seeded;" the small size tends to lead one to *Crotonopsis* in the key under the choice "Fruit 1-locular, 1-seeded, indehiscent."

664 In description of *Crotonopsis elliptica*, line 2, "3 dm" should be "3 cm."

668 In description of *Euphorbia*, line 1, delete comma after "laticiferous."

669 In *Euphorbia* key the choice leading to *E. purpurea* is improperly indented. It should be aligned with the choice immediately above it, which begins "Stem leaves less than 5 cm long."

672 Stems of *Euphorbia corollata* (in AL) are commonly 4 dm tall and range up to 1 m.

675 In Buxaceae key, line 2, "Inflorescenses" should be "Inflorescences."

676 In description of *Rhus vernix*, line 5, "Panicle ... 1–2 cm long" should be "Panicle ... 1–2 dm long."

678 Specimens of *Rhus glabra* have up to 25 leaflets.

679 In *Ilex* key I. Fruiting Specimens, the choices at the couplet "Sepals glabrous, non-ciliate" vs. "Sepals pubescent or ciliate" are reversed. *Ilex laevigata* description, p. 684, states "Sepals not ciliate," whereas *I. verticillata* description, p. 683, states "sepals ciliate." The key leads are reversed.

679 In *Ilex* key II. Flowering Specimens, the choices leading to *I. decidua* and *I. ambigua* indicate flowers 4-merous and 5-merous, respectively. Descriptions on p. 681 and 683 indicate the reverse is true for these species. Exchange the numbers "4" and "5" in the key.

681 In description of *Ilex opaca*, line 7, concerning "Pistillate flowers," the entry "stamens 4" should be "carpels 4."

685 In the figure title for *Paxistima canbyi*, the generic name is incorrectly spelled (in the manner noted in the description on the same page).

690 In descriptions of *Aesculus sylvatica*, lines 2–4, and *A. octandra*, line 2, add a hyphen to "non glandular" as shown in *A. pavia* description, line 2.

693 In Rhamnaceae key, the choice leading to *Hovenia* states "leaves lanceolate to oblanceolate" vs. "leaves elliptic or weakly obovate" in the choice leading to *Rhamnus*. Description of *H. dulcis* on p. 694 states "Leaves ... ovate to elliptic."

694 In Vitaceae key, line 2, and *Ampelopsis* key, p. 697, line 1, add "pinnately" after "ternately."

694 In Vitaceae key, change the choice leading to *Vitis* to "pith in mature stems brown, interrupted ..." and the choice leading to *Ampelopsis* to "pith in mature stems white, usually continuous ..."

694 Drupes of *Parthenocissus quinquefolia* have as many as 4 seeds.

702 In *Sida* key, line 1, delete "about 3–6 mm wide" and change to "Leaves linear; petals 1.5 cm or more long." This makes key more consistent with description of *S. elliottii*.

702 period after "peduncles 0.7–3 cm long" to a comma.

710 In *Hypericum* key, "capsule 1 celled" in the first choice leading to *H. cistifolium* should be "capsule 1-celled."

710 In *Hypericum* key, *H. mutilum* is reached via the choice "Leaves more than 3 mm wide, 5–7 nerved." Description, p. 715, states "Leaves ... 3–5 nerved." Change the minimum nerve number in the key to "3."

715 Descriptions of *Hypericum* ssp. 22–24 state "stamens few." Numerical limits of the term "few" are not set anywhere in the manual, but in *Hypericum* it could mean no fewer than "5" according to the generic description, p. 709.

724 In the Violaceae key, the choice leading to *Viola lanceolata* should be changed to "Leaves narrowly lanceolate or oblanceolate, tapering to long acute at base."

725 Key choice leading to *Viola papilionacea* states "marginal teeth distinct, 0.5–1 mm long on the shorter side of the tooth." Description states "teeth 2–7 mm long, 0.5–2 mm wide." Length and width of teeth seem to be reversed in the description, and neither agrees with the key; length of a tooth that is asymmetrical should be measured either according to the distinct sides or from a line connecting adjacent sinuses to the apex. The latter measurement would be actual length of the tooth.

729 Authority for *Viola septemloba* should be Le Conte rather than House.

731 In description of *Viola canadensis*, line 6, "sepals 4–10 cm long" should be "sepals 4–10 mm long."

734 Description of Passifloraceae states "ovary 3-locular." Change this to "ovary 1-locular, with 3 parietal placentae."

736 Description of *Elaeagnus* states "Fruit drupaceous," which is incorrect. The fruit is an achene encased by a fleshy calyx, and thus it superficially resembles a drupe.

743 In description of *Rhexia lutea*, line 2, "Leaves ... to 2 8 cm long" should be "Leaves ... to 2.8 cm long."

744 In Onagraceae key, to the choice "Seeds not comose, petals yellow" add the words "or absent." Apetalous species of *Ludwigia* more readily fit the key after this change.

744 Leaves of some specimens of *Ludwigia leptocarpa* are decurrent; this key characteristic leads to *L. decurrens* unless capsule shape is used as the primary distinguishing feature.

750 In *Oenothera* key, the first choice states "Capsule ... not angled." Description of *O. rhombipetala*, which can be reached only via this choice, in line 6 states "Capsules ... angled." Change the primary couplet to "Capsules ... not or only slightly angled" vs. "Capsules ... usually sharply angled or winged."

750 Specimens of *Oenothera biennis* (in AL) produce sharply angled (prismatic) seeds that are obliquely angled at the ends. Ovules and seeds appear to be ascending in the locule rather than "horizontal" as stated in the *Oenothera* key. Also, the description of *O. rhombipetala*, to which one is led if features of *O. biennis* are used to key a specimen, states "seeds not seen."

750 In *Oenothera* key, the choice leading to *O. humifusa* states "Plant prostrate." Stems of this species may be erect or ascending with divergent branches that tend to be depressed at the ends, but only rarely are they completely prostrate. Change the key to "Plant usually ascending, becoming prostrate distally."

758 Key choice in Araliaceae that leads to *Aralia* should be changed to "Herb, shrub, or small tree; leaves compound." This will permit large plants of *A. spinosa* to be identified more easily and also cause key and description to agree.

760 In description of *Panax*, line 3, "flowers polygamo-dioecious" should be "flowers perfect or imperfect." The terms dioecious, monoecious, polygamous, and their combined forms properly apply to species or plants in a collective sense (as used for Araliaceae on p. 758), but not to an individual plant nor to flowers. It is simply not possible for one plant to be dioecious, nor is it appropriate to describe flowers as polygamous.

760 "*Panax quinquefolium*" should be "*Panax quinquefolius* L.," and "*P. trifolium* L." should be "*P. trifolius* L." The genus *Panax* is masculine in gender.

761 In description of Apiaceae, line 9, "obsurely" should be "obscurely."

761 In Apiaceae key, line 8, "simpler" should be "simple."

763 In *Hydrocotyle* key, the terms "pedunculate" and "sessile" are reversed in the couplet leading to *H. americana* and *H. sibthorpioides* according to the descriptions on p. 765 and figures on p. 764.

778 In *Apium* key, "3s" should be "3's."

781 In description of *Conium maculatum*, last line, "poison" should be "poisonous."

783 In description of *Coriandrium*, delete apostrophes at ends of lines 5–7, and add period at end of line 8.

786 Choices in couplet leading to *Angelica triquinata* and *A. atropurpurea*, respectively, state "seed adherent to pericarp" vs. "seed loose in pericarp." Descriptions of these species indicate the opposite conditions except that "adnate" is substituted for "adherent." The descriptions are correct, and the key should be changed.

786 Add asterisk(*) after *Angelica triquinata* Michaux; the species is illustrated on p. 785.

789 In *Nyssa* key, line 1, "2s or 3s" should be "2's or 3's."

792 In description of *Clethra acuminata*, leaf base is stated to be "cuneate." It is commonly rounded or truncate. Leaf surface is described as "essentially glabrous," but leaves are pale beneath due to stellate trichomes.

793 In description of Ericaceae, line 5, change to "Fruit a capsule, berry, or berry-like drupe."

793 In Ericaceae key, line 5, "Evergeeen" should be "Evergreen."

793 In Ericaceae key, line 6, add "or tree" after shrub. This causes *Oxydendrum* on p. 808 to fit this choice and facilitates its identification.

793 In final couplet in Ericaceae key leading to *Gaylussacia* and *Vaccinium*, respectively, change first choice to "fruit a fleshy drupe containing 10 pyrenes" and the second to "fruit a berry, usually more than 10-seeded (rarely fewer than 10-seeded)."

797 In *Rhododendron* key I, Flowering Specimens, *R. canescens* can be reached only under the choice "Pedicels and calyx non-glandular." Description on p. 800 states "Calyx lobes ... glandular ciliate."

798 In description of *Rhododendron minus*, line 6, add "or orange" after the phrase "spotted with green."

803 Specimens of *Kalmia latifolia* occasionally may have flowers in simple corymbs as well as "compound corymbs" as stated in the description.

813 Specimens of *Vaccinium elliottii* (in AL) sometimes flower before the leaves develop; absence of leaf characteristics in such plants causes them to be difficult to key.

814 In *Vaccinium* key, line 15 exactly repeats line 14, "Leaves less than 1.5 cm long, thick, margins revolute." Delete one of these lines.

821 In *Lysimachia* key, in the first half of the couplet beginning with "Leaves rounded to petiole" vs. "Leaves with long cuneate base," add "or widely cuneate" before the semicolon.

825 In last sentence of *Limonium* description, compounded from two independent clauses, either change comma to semicolon after "understood" or else add "and" before "more" if the comma is kept.

826 Description of *Diospyros virginiana* states "corolla yellow." Change to "corolla creamy white to pale yellow." Corollas are more white than yellow in fresh flowers but darken during drying and may appear in dried specimens to have been yellow before being pressed.

827 Description of *Halesia*, the key, and species descriptions use fruit characteristics as important taxonomic criteria, but the fruit type is not stated. Change generic description, line 6, to "Fruit a dry, winged, beaked, corky drupe, 1–3 seeded."

827 In *Halesia* key, change choice for *H. carolina* to "Fruit equally 4-winged," and add "or with 2 prominent wings and 2 smaller ones" to the lead for *H. diptera*.

827 In description of *Styrax*, line 1, delete comma after "axillary."

829 In Oleaceae key, line 1, change "Leave" to "Leaves."

832 In *Ligustrum* key, *L. ovalifolium* can be reached only under the choice "Branchlets pubescent." Description states "Branchlets and petioles glabrous." The description is correct.

832 Description of *Lantana* states "Calyx ... lobes less than 1 mm long, tube 1.5–2 mm long." Choice leading to *L. camara* in key states "Calyx lobes acute, as long as or longer than the tube."

832 Description of Loganiaceae, line 4, states "ovary superior." Appropriate leads for identification of several genera in this family are not presented under this choice in Key 7, pp. xxviii–xxxii, but Loganiaceae is easily reached when one incorrectly follows the choice "ovary inferior" on p. xxx. Extra leads to assist in correct identification despite misinterpretation of a feature are useful in a key, but for success to depend on such misinterpretation is not.

839 Add asterisk(*) after *Sabatia dodecandra* var. *coriacea* (Ell.) Ahles; the species is illustrated on p. 837.

839 Descriptions of *Gentiana saponaria* and *G. catesbaei*, p. 842, give lengths for corolla pleats that appear to be inconsistent with the key. Change information concerning lobes in the final couplet in the *Gentiana* key to "pleats about 3/4 as long as the lobes" vs. "pleats about 1/2 as long as the lobes" for these species, respectively.

842 *Frasera* is preferred over *Swertia* by most authors.

845 Description of Apocynaceae, line 2, states "Flowers 5-merous." Figure of *Amsonia tabernaemontana* on p. 846 shows two flowers with 6 petals each. No mention is made under *Amsonia* of the occurrence of such 6-merous corollas; thus, it is not clear whether such flowers exist or not, and the figure is confusing.

845 The asterisk after *Amsonia tabernaemontana* var. *tabernaemontana* is out of place by a couple of spaces.

845 Specimens of *Amsonia ciliata* (in AL) sometimes have glabrous stems. The key could be changed to "stems usually pubescent" in the choice leading to this species.

848 In description of *Asclepias*, line 4, and the key that follows, the term "gynostegium" is used as a major diagnostic characteristic. This term is briefly described in the family description on p. 847–848 but not in the glossary on p. li.

848 First choice in *Asclepias* key I. Flowering Specimens should be changed to "Petals white, or tinged with green or rose and the leaves linear."

848 Second line in *Asclepias* key I should be changed by adding "to narrowly lanceolate" after "linear."

858 Key choice leading to *Cuscuta compacta* states "Flower sessile, subtended by 3–5 fringed bracts." Description on p. 859 states "Flowers ... subtended by 2–5 ... bracts." The correct minimum number of bracts appears to be 3.

858-861 Use of corolla tube length for "flower" length in descriptions of *Cuscuta* species, as stated in genus description, "often makes the corolla size larger" (than the flower). This is confusing and clearly unnecessary. The data should have been given as corolla length, tube length, and lobe length, respectively.

861 *Bonamia* spp. are usually treated under synonyms in *Stylisma*, some of which are correctly indicated at the ends of descriptions, pp. 861–863 (Myint 1966).

861 Add an asterisk (*) after *Bonamia aquatica* (Walter) Gray; the species is illustrated on p. 862. The correct author citation for *Stylisma aquatica* is (Walter) Raf.

862 In figure of *Jacquemontia tamnifolia*, the apex downward orientation of stem and contorted leaf blade and peduncle make it appear that the inflorescence originates from the abaxial base of the petiole rather than the axil. The figure is confusing when compared with a typical plant.

863 Add an asterisk(*) after *Bonamia humistrata* (Walter) Gray; the species is illustrated on p. 862.

863 *Bonamia patens* (Desr.) Shinners = *Stylisma patens* (Desr.) Myint.

863 *Bonamia patens* var. *angustifolia* (Nash) Shinners = *Stylisma patens* var. *angustifolia* (Nash) Myint.

863 In description of *Calystegia*, line 4, "often" should be capitalized.

864 Key choice leading to *Ipomoea quamoclit* and *I. coccinea* states "Corolla ... scarlet." Descriptions of both species on p. 866 state "corolla crimson." Both terms denote deep red, but "crimson" has the connotation of blood red, whereas "scarlet" suggests orange red. The key is more precise in this case.

869 Key choice leading to *Phlox pilosa* states "Sterile shoots absent." Description on p. 870 also states this but adds "or not persistent." This information would be helpful in the key.

874 In description of Hydrophyllaceae, line 2, "scorpioid" should be "helicoid."

878 In description of Boraginaceae, line 2, "helicoid cymes of false racemes or spikes" should be "helicoid cymes or false racemes or false spikes (cincinni)." Any reference to inflorescences in this family as "racemes" or "spikes," unless prefaced by "false," or as "scorpioid cymes" is erroneous and should be corrected.

879 In description of *Heliotropium*, line 3, "helicoid false racemes" should be "helicoid false spikes (cincinni)."

879 Ditto, *Heliotropium* key and various species descriptions.

879 In key choice leading to *Heliotropium amplexicaule*, "sub-sessile" should be "subsessile."

879 In description of *Heliotropium indicum*, line 3, "false, helicoid raceme" should be "helicoid false spike" or simply "cincinnus."

880 In description of *Heliotropium amplexicaule*, line 6, delete comma from "two, 2-seeded mericarps."

880 In descriptions of *Cynoglossum*, line 2, and *Hackelia*, line 3, the term "sympodial" describing "false racemes" is redundant. A sympodial axis (or rachis) causes a helicoid or scor-pioid cyme to be a false raceme if the flowers are pedicelled or a false spike (e.g., a cincinnus, when axis is helicoid) if the flowers are sessile or subsessile.

882 In description of *Echium*, line 4, "scorpioid" should be "helicoid."

884 In description of *Myosotis*, line 1, delete comma after "helicoid."

884 Key choice leading to *Myosotis laxa* states "Stems usually lax and decumbent, neither creeping nor stoloniferous." Description states "stems slender, decumbent or spreading, rooting at the nodes." The stem condition described can hardly be distinguished from one that is creeping and/or stoloniferous.

887 In description of *Plaqiobothrys*, line 2, "Spikes" should be "Cincinni."

888 Key choice leading to *Verbena bracteata* states "Subtending bracts 3x or more as long as calyx." Description on p. 891 states "bracts 1–2x length of calyx in flower, usually longer in fruit." In a flowering specimen there is no overlap between limits of this feature as stated in the key and in the description.

888 In *Verbena* key, line 9, delete dotted line and "3. *V. officinalis*" following the choice "Plant annual; corolla about 3 mm broad."

888 Key choice leading to *Verbena brasiliensis* states "mature central spike less than 3 cm long." Description states "Spikes ... 0.5–4 cm long." Change upper limit in key to "4 cm."

892 In description of *Lippia*, line 4, "spike 8–15 cm long" should be "spike 8–15 mm long."

892 In description of *Lippia lanceolata*, line 4, add one bracket before "SE."

894 In description of *Callicarpa*, line 3, add "4- or 5-merous," after "Flowers;" line 4, delete "5-" from "5-toothed;" line 5, delete "5," from "lobes 5, spreading."

895 In description of *Phryma leptostachya*, line 9, "uncinnate" should be "uncinate."

898 In some specimens of *Hyptis alata* the lower peduncles are shorter than the ones above them on the stem.

898 Description of *Hyptis alata*, lines 2–3, states "Leaves ... 5–15 cm long." Figure on p. 899 indicates dimensions that convert to leaf lengths well below the minimum stated.

908 In description of *Leonotis nepetaefolia*, line 4, delete comma after "1–3."

909 Illustration labeled "*Stachys floridana*" is actually *Stachys tenuifolia*, p. 911.

910 In description of *Lamium purpureum*, line 1, "1–1.5 mm" should be "1–1.5 cm."

910 Delete asterisk(*) after 1. *S. floridana* Shuttlew. and place it after 6. *S. tenuifolia* Willd. on p. 911.

922 In description of *Lycopus amplectens*, line 6, "barley" should be "barely."

930 In *Solanum* key, line 11, "2.8 cm wide" should be "2–8 cm wide."

930 In *Solanum* key, in the choice leading to *S. tuberosum*, "pinnatified" should be "pinnatifid."

932 *Solanum torreyi* Gray = *S. dimidiatum* Raf.

932 In description of *Solanum tuberosum*, line 3, "unequaly" should be "unequally."

933 In descriptions of *Solanum americanum*, line 10, and *Datura*, line 5, "poison" should be "poisonous."

936 In Scrophulariaceae Key 2, choice leading to *Penstemon* states "pubescence not glandular." Descriptions of spp. 2–6 on pp. 947–949 indicate glandular pubescence in the inflorescence. The key is misleading on this point.

936 In Scrophulariaceae Key 3, second line from bottom, "antinomorphic" should be "actinomorphic."

943 In description of *Mimulus*, line 1, the period after "punctate" is in the position of a hyphen.

944 Figure labeled "*Mimulus alatus*" should be labeled "*Mimulus ringens*," and move the asterisk(*) from the former name to the latter on p. 943.

945 In key to *Chelone*, line 5, "*C. lyoni*" should be "*C. lyonii*."

956 Key choice leading to *Aureolaria flava* states "Leaves elliptic ovate" vs. "Leaves lanceolate" for *A. laevigata*. Description of the former on p. 958 states "Leaves lanceolate or elliptic-lanceolate." Thus, the primary diagnostic feature in the key does not agree with the description.

958 Specimens of *Agalinis fasciculata* (from AL) often have pedicels as long as or slightly longer than the calyx tube.

960 *Agalinis setacea* (J. F. Gmelin) Raf. should be *A. sericea* (Walter) Raf.

961 In description of *Agalinis decemloba*, line 4, the first letter in "localities" is below the line.

964 In *Catalpa* key, corolla dimensions refer to "broad," whereas in the species descriptions corolla dimensions refer to "long." Perhaps both length and width should be indicated in both places.

965 In description of *Conopholis americana*, line 4, the word "to" after "than" should be deleted.

966 Description of Lentibulariaceae states "annual or perennial ... herbs," but growth form is described only for the genus *Pinguicula* and not *Utricularia* within the family treatment.

971 In Acanthaceae key, change second couplet to "Corolla nearly actinomorphic, lobes 5, subequal; stamens 4" vs. "Corolla zygomorphic, strongly bilabiate; stamens 2."

972 In *Ruellia* key, change the second half of the first couplet to "Flowers in open cymes, or solitary or in glomerules (the latter sessile or subsessile) at upper and middle nodes."

975 In *Plantago* key, *P. lanceolata* is available only under the choice "Leaves 1 cm or less wide" on p. 974. Description states "Leaves 0.7–5 cm wide." Thus, it is rarely possible to use the key successfully with this species. At the end of the key add: "Plant annual; bracts obtuse" vs. "Plant perennial; bracts acute or acuminate" to distinguish 8. *P. virginica* and 4. *P. lanceolata*, respectively.

977 In description of *Plantago aristata*, line 4, "solid" should be "hollow."

978 In description of Rubiaceae, line 5, add "schizocarp" after "berry." The fruit is a schizocarp in *Diodia*, *Spermacoce*, and *Richardia*, and appropriate annotation of the descriptions on pp. 979–981 is needed.

979 In description of *Diodia teres*, line 3, add "or pink" after "white."

979 In descriptions of *Spermacoce*, line 5, and *Richardia*, p. 981, line 2, "carpels" should be "segments." The number of segments in the latter is more often 3 than "4" as stated.

984 In *Galium* key, line 7 from bottom, "but now downy" should be "but not downy."

988 In Caprifoliaceae key, the choice leading to *Lonicera* states for species with axillary flowers "ovaries fused at the base." Descriptions of species reached via this same information in the *Lonicera* key, p. 990, state "ovaries separate to near base" for *L.* x *bella* and *L. japonica* and "ovaries separate to base" for *L. fragrantissima*. The keys are misleading for these three species and correct only for *L. canadensis*.

990 In description of *Lonicera japonica*, line 2, "evergreen" begins with a small capital "E."

993 In key choice leading to *Viburnum rafinesquianum*, delete comma between "leaves" and "sessile."

994 Peduncles of *Viburnum prunifolium* (in AL) are as long as 1 cm.

995 In description of *Viburnum dentatum* var. *dentatum*, line 2, "fosests" should be "forests."

1002 In the Campanulaceae key, line 6, add "often appearing as a leafy spike" after "axillary."

1002 Key to *Specularia* species and descriptions for *S. perfoliata* and *S. biflora*, respectively, state "pores of capsule at or below middle" vs. "pores of capsule near the top" for these taxa. Figures of capsules for these species on p. 1003 are reversed.

1005 In description of *Sphenoclea zeylandica*, line 8, the bracket after "Miss." is backward.

1005 In *Lobelia* key, line 15, "8 mm" should be "7 mm." Descriptions of *L. amoena* and *L. puberula* on p. 1007 give minimum corolla tube lengths of 7 mm.

1007 In description of *Lobelia siphilitica*, top line, "12–15 cm" should be "12–15 mm."

1009 In description of Asteraceae, line 6, "plant is" should be "plants are." See annotation concerning *Panax*, p. 760.

1009 In Asteraceae key, the choice leading to Key 3 states "Leaves dissected or lobed." Description of *Berlandiera*, p. 1105, which is available only through Key 3, states "Leaves ovate to lance-ovate ... crenate." *Berlandiera* should be included in Key 5.

1009 In Asteraceae key, the genus *Chondrophora* is included only in Key 4, which is available only under the choice "Heads of ray and disc flowers." Description of *C. nudata* (Michaux) Britton, p. 1098, states "Flowers all discoid, yellow." The description is correct, and this genus should be available in both Key 9 and Key 10 for Asteraceae. See also pp. 1010 and 1098 for other notes concerning *Chondrophora*.

1009 In Asteraceae key, the first choice leading to Key 7 states "Leaves alternate." Description of *Melanthera*, p. 1120, which is available only in Key 7, states "Leaves opposite." *Melanthera* should be included in Key 10.

1009 In Asteraceae key, species with simple, alternate leaves and heads of perfect disc flowers only should key out in Key 9. Description of *Gaillardia aestivalis* states "Rays usually absent," but *Gaillardia* is available only in Key 5 which is reached only under "Rays present."

1009 In Asteraceae key, scapose plants of *Crepis japonica* lead to Key 11 which does not include *Crepis*.

1009 In Asteraceae key, in the key choice leading to Key 12, add a comma after "bracts."

1010 In Asteraceae Key 1, line 2, "Involucre without forked trichomes" should be changed to "Involucre glabrous or with simple trichomes."

1010 Key choice leading to *Prenanthes* in Asteraceae Key 1 requires a decision that rays are "white or cream-colored" vs. "yellow" for *Crepis*. *Prenanthes* spp. 3–5, pp. 1021-1022, are described as "yellowish," "yellow," or "straw-yellow," respectively. The key is misleading about this point.

1010 In Asteraceae Key 2, the first choice leading to *Chrysanthemum* falls under "Rays 1 cm or more long." Description of *C. parthenium* on p. 1137, which should be keyed via that choice, states "Rays ... 0.5–1 cm long."

1010 As stated above for p. 1009, *Chondrophora* should not appear in Key 4, but the key choice states "stem leaves less than 2 cm long and 3 mm wide" for this genus vs. "stem leaves larger" for *Solidago*. Description of *Chondrophora*, p. 1098, states "Stem leaves ... 5–15 cm long and 3–10 mm wide." Delete reference to width in *Chondrophora* half of couplet and change "larger" to "longer" in the *Solidago* half.

1011 Key choice in Asteraceae Key 4 leading to *Heterotheca* states "basal leaves less than 1 dm long." Descriptions of *Heterotheca* ssp. 1–4 indicate "Basal leaves to 4 dm long" and a minimum length of "1 dm." Only Asteraceae Key 4 provides a lead to *Heterotheca*, and leaf lengths of the key and certain species descriptions are mutually exclusive.

1011 In Asteraceae Key 6, plants of *Erigeron* with bracts in as many as 3 imbricate series lead to *Solidago* or *Aster*.

1012 In Asteraceae Key 7, the "t" in "spine-tipped" is not clearly printed.

1013 In Asteraceae Key 9, lines 24 and 25, "and" is repeated. Delete one "and."

1013 Inflorescence form in *Liatris* sometimes appears paniculate due either to injury near the stem apex or else the pattern of flowering from the apex downward may allow some lower axillary branches to develop. Such specimens lead one past the correct generic name in Key 9.

1013 In Asteraceae Key 9, specimens of *Marshallia* with solitary heads (a common condition in *M. obovata*) lead one to *Liatris*. To the choice leading to *Liatris*, add "on short peduncles" after "solitary," and change the other half of this couplet to "Heads corymbose, paniculate, thyrsoid, or solitary on long peduncles."

1013 In Asteraceae Key 9, choices leading to *Vernonia* require either "Flowering stem with a basal rosette" or "Outer bracts long acuminate or acute and the pappus purple." Description of "*V. noveboracensis*" only among the species with "basal leaves absent" has pappus described as "dark brown to purple." Pappus of others is described as "whitish, tawny, pinkish or purple" for *V. angustifolia*, "yellowish to tan" for *V. glauca*, and "tawny to purple" for *V. altissima*. Key 9 is very weak at the point at which a decision for or against *Vernonia* must be made.

1013 In Asteraceae Key 9, couplet leading to *Erigeron* near end of key compares primarily leaf shape "linear to linear-lanceolate" with leaf size "more than 1 cm wide" for the remainder of the key. Delete reference to shape from the first choice and change to "Leaves usually less than 1 cm wide, conspicuously ciliate."

1014 In Asteraceae Key 10, add a comma after "brownish" in the first line that leads to *Bidens* on this page.

1014 In Asteraceae Key 12, change first couplet leading to *Coreopsis* and *Bidens*, respectively, to "Nutlets winged, or not awned" vs. "Nutlets wingless, awned." *Coreopsis* description, p. 1120–1121, states "Nutlets ... winged or wingless," but specimens of *C. tinctoria* lead one to *Bidens* when absence of wings on nutlets serves as the sole distinguishing feature.

1014 Second part of Asteraceae Key 12 is misleading for *Coreopsis* specimens (e.g., *C. auriculata*) with entire leaves and rays with prominent teeth. The second half of couplet separating *Chrysogonum* from *Coreopsis* and *Bidens* should be changed to "Rays entire or more than 3-toothed at apex; leaves entire or serrate."

1015 In Asteraceae Key 13, lines 6 and 7 are the same. Change the second half of this couplet to "Leaves not lobed or dissected; disc flowers fertile."

1025 The key to *Hieraceum* species is misleading about corymb shape under the choice "Heads corymbose" because 5. *H. venosum*, p. 1026, with a "flat-topped corymb" is reached via the choice "Heads in loose concave or convex corymbs; ..." Changing the "or" to "to" improves the wording here. The real contrast is between "compact" and "loose" at this point.

1027 In description of *Hieraceum gronovii*, line 4, add "corymbiform" before "panicle."

1027 Choice leading to *Crepis japonica* states "Involucre 4–5 mm long." Description on p. 1028 states "Involucres 5–6 mm long." Sizes in the key seem to be accurate in this case.

1027 In key to *Crepis*, line 6, "Infolucre" should be "Involucre."

1028 In *Hypochoeris* key, line 2, "bracteate" should be "bracteal."

1031 Key choice leading to *Krigia biflora* and *K. montana* states "involucre 9 mm or more long." Descriptions of these species state "Involucres ... 7–10 mm long" and "7–9 mm long," respectively. Based upon species descriptions, the *Krigia* key can be improved by changing "8 mm" to "5 mm" in the choice leading to *K. oppositifolia* and "9 mm" to "7 mm" in the contrasting choice discussed above.

1034 Heads of *Cacalia atriplicifolia* (in AL) are 3–7 flowered.

1035 In second half of couplet separating *Senecio obovatus*, plants are described as "caespitose." Spelling of this term is not consistent with the Glossary, p. xlvii, nor the description of *S. millefolium* on p. 1035. Both forms of the term are correct, but consistency in spelling such terms is needed. "Cespitose" is shorter, also.

1037 *Senecio smallii* Britton = *S. anonymus* Wood.

1041 In description of *Carduus lanceolatus*, line 2, "aracchnoid" should be "arachnoid."

1047 In *Elephantopus* key, line 5, add "usually" after "leaf" and before "7 cm."

1054 In *Eupatorium* key, period at end of line in choice "Leaves opposite ...; heads." should be a comma.

1055 In *Eupatorium* key, line 16, "distincly" should be "distinctly."

1055 Lower portion of *Eupatorium* key starting with "Leaves sessile, or petioles less than 5 mm long" is not properly indented. This choice should be aligned with the line at the top of the page concerning "Leaves with petioles ..."

1056 Height of *Eupatorium fistulosum* plants commonly approaches 4 m; this is much taller than the "2.5 m" given for the "similar" *E. maculatum.*

1072 In *Aster* key, line 12, the choice leading to *A. linearifolius* should be changed to "Leaves less than 4 mm wide, usually less than 3 cm long" to agree with limits in description on p. 1081.

1072 In *Aster* key, the first lead to *A. paludosus* under "Leaves entire" appears only under "Stems pubescent." Description on p. 1077 states "stems glabrous to pubescent."

1073 In *Aster* key, the first choice leading to *A. pilosus* is under "Leaves not linear, not fleshy." Description on p. 1081 states "Leaves elliptic to linear." This inconsistency can be corrected by adding "or" before "not fleshy."

1073 In *Aster* key, near the end under choice "Involucres less than 1 cm long" before the couplet leading to *A. linearifolius* and *A. squarrosus*, the "r" in "rigid" is missing.

1074 In description of *Aster paternus*, line 6, "sericeus" should be "sericeous."

1074 In description of *Aster commixtus*, line 5, "gradulaly" should be "gradually."

1076 In description of *Aster undulatus*, line 6, add a comma after the second "auriculate" (before "widely winged").

1080 Involucres in fresh specimens of *Aster carolinianus* often measure no more than 8 mm broad; in dried specimens the breadth approaches the "15–18 mm" stated in the description, primarily due to a change of shape.

1082 In description of *Aster racemosus*, line 1, "ro" should be "or."

1085 Key choice leading to *Solidago sempervirens* states "Rhizomes absent." Description on p. 1091 states "rhizomes short and stocky." Change couplet to "Rhizomes elongate" for 13. *S. stricta* and "Rhizomes short" for 14. *S. sempervirens.*

1086 In *Solidago* key, couplet on lower page "Widest leaves 7 mm or less wide" vs. "Widest leaves 10 mm or more wide" is inconsistent with descriptions of both species under the first choice and *S. radula* under the second. This couplet is improved by reliance on the qualitative leaf serration data given and adding "usually" before the size limits.

1087 Key choice leading to *Solidago nemoralis* states "Leaves without prominent lateral veins." Description on p. 1092 states "lateral veins evident." Under the contrasting choice "Leaves with a pair of prominent lateral veins," the lead to *S. altissima* states "veins obscure." Description of the latter on p. 1095 states "lateral veins evident" just as for *S. nemoralis.* Clearly, either key or descriptions or both need modification. Add "not" after "veins" in *S. nemoralis* description. Delete reference to veins in couplet leading to *S. radula* and *S. altissima*, since it is not clear whether reference is made to lateral veins or netted veins.

1093 Figure of *Solidago radula* shows a panicle much more elongate than described on p. 1092 as "about as broad as long."

1098 Change stem leaf length from "5–15 cm" to "5–15 mm" in the *Chondrophora* description; perhaps the description should treat the basal leaves, the dimensions of which seem to be presented for "stem" leaves. The correct name for the species is *Bigelowia nudata* (Michaux) DC.

1103 In description of *Silphium*, lines 2 and 3, "Rays 3, notched at apex" should be "Rays 3-notched at apex."

1103 In key to *Silphium* species, "of" should be "or" in the choice leading to 5. *S. asteriscus.*

1103 In description of *Silphium terebinthinaceum*, line 3, the final numeral in "1–1.] cm long" appears to be a small bracket. Change to "1–1.2 cm long."

1108 Key choice leading to *Rudbeckia fulgida* states "Chaff obtuse." Description states "Chaff acute." Description should be changed to "Chaff mostly obtuse, rarely acute," but this does not fully clear up inconsistency with the key.

1111 In description of *Helianthus*, line 5, after "pappus scales 2" add "or sometimes 3 on ray flowers."

1111 In *Helianthus* key, in line 2 of the first choice place a comma after "width" and add "less than half as wide as basal leaves."

1113 In description of *Helianthus angustifolius*, line 3, "3–15 cm wide" should be "3–15 mm wide."

1118 In *Verbesina* key, the choice leading to *V. occidentalis* states "Petioles not winged at base." Description states "petioles winged, the tissue decurrent on the stem as a wide wing" and the figure on p. 1119 shows this. Change "winged" to "auricled" in both choices in the key.

1121 In *Coreopsis* key, 10th line from the end, change the colon after "not stoloniferous" to a semicolon.

1126 In description of *Bidens* is the statement "No single character will separate this genus from *Coreopsis*." In Asteraceae Key 12, p. 1014, single characters are used in two places to separate these genera; see appropriate annotations for p. 1014.

1129 In description of *Bidens pilosa*, line 7, delete brackets around "absent."

1129 In *Marshallia* key, line 2, "stem leaves ... bracteate" should be "stem leaves ... bracteal."

1129 In *Marshallia* key, all the species under "corolla white" are described on p. 1131 as having "Corollas pink" except *M. obovata*. The latter description states "Corolla white, rarely pinkish." Either add "or pink" after "corolla white" or delete reference to corolla color in the primary couplet.

1134 In description of *Hymenopappus scabiosaeus*, lines 3 and 4, descriptions of basal and cauline leaves are switched. Basal leaves are bipinnately dissected and cauline leaves are once pinnate.

1142 "Betcke" is out of alphabetical order; it should appear before "Beyrich."

1142 "Bonpland" should appear after "Boivin."

1142 "B.& H." should appear after "Austin" on p. 1141.

1146 "Garden," authority for *Stillingia sylvatica*, p. 667, should be added to author list: "Alexander Garden, 1730-1791."

1151 "Rudge," authority for *Carex intumescens*, p. 254, should be added to the author list: "Edward Rudge, 1763-1846."

1154 "Turzc." should be "Turcz."

1157 For *Achyranthes*, "412" should be "422."

1157 "Amaranthaceae" should appear after "amaranth."

1158 "*Amphicarpa*" should be "*Amphicarpaea*."

1158 For *Anoda cristata* var. *brachyanthera*, "720" should be "702."

1159 For *Bacopa monnieri*, "928" should be "938."

1159 "Blue Curls," common name for *Trichostema*, p. 897, should be listed in index.

1162 For *Chrysopsis*, "1110" should be "1101."

1165 "Dandelion" should appear after "Dame's Rocket."

1165 Index words starting with "Dog" are not alphabetized.

1165 "Duck-potato" should appear after "*Duchesnea indica*."

1165 "*Ellisea*" should be "*Ellisia*."

1169 For *Kneiffia fruticosa*, "652" should be "752."

1169 "*Koellia hyssopifolia*" should appear after "*Koellia hugeri*."

1170 For *Leontodon nudicaulis*, "1082" should be "1028."

1170 "*Lindera melissaefolium*" should be "*Lindera melissifolia*."

1171 "*Ludwigia sphaerocarpa*" should appear after "*Ludwigia spathulata*."

1172 "Mullein, Woolly, 945" should be included in index.

1173 "*Panax quinquefolium*" and "*Panax trifolium*" should be "*Panax quinquefolius*" and "*Panax trifolius*," respectively.

1174 For *Phalaris*, "12" should be "121." This deletion (or loss) of the numeral appeared in the Sixth and subsequent printings, apparently not in earlier ones.

1176 For *Quercus falcata*, add comma after "380."

1176 *Quercus falcata* var. "*pagodifloia*" should be "*pagodifolia*."

1176 For *Quercus georgiana*, add comma after "378."

1177 "Rockrose" should appear before "Rocky."

1180 "*Sygonanthus*" should be "*Syngonanthus*."

1182 For *Vaccinium atrococcum*, "186" should be "816."

1182 *Vaccinium stamineum* var. "*melanocarp um*" should be "*melanocarpum*."

References

Correll, D.S. 1972. Manual of the vascular plants of Texas: I. Additions and corrections. Amer. Midl. Naturalist 88:490–496.

Cronquist, A.J. 1980. Vascular flora of the southeastern United States. Vol. 1, Asteraceae. Univ. North Carolina, Chapel Hill. 261 pp.

Freeman, J.D. 1975. Revision of *Trillium* subgenus *Phyllantherum* (Liliaceae). Brittonia 27:1–62.

Hill, S.R. 1981. Supplement to flora of the Texas Coastal Bend by F.B. Jones. Sida 9:43-54.

Hitchcock, A.S. 1950. Manual of the grasses of the United States, 2nd Ed., rev. by A. Chase. U.S. Government Printing Office, Washington. 1051 pp.

Isely, D. 1990. Vascular flora of the southeastern United States. Vol. 3, Part 2, Leguminosae (Fabaceae). Univ. North Carolina, Chapel Hill. 258 pp.

Johnston, M.C. 1990. The vascular plants of Texas: A List Up-dating the Manual of the Vascular Plants of Texas, 2nd Ed. Publ. by the author, Austin. 107 pp.

Kartesz, J.T. 1994. A synonymized checklist of the vascular flora of the United States, Canada, and Greenland, 2nd Ed. Timber Press, Portland, OR. Vol. 1, Checklist, 688 pp.; Vol. 2, Thesaurus, 816 pp.

Lawrence, G.H.M. 1951. Taxonomy of vascular plants. Macmillan, New York. 823 pp.

Morin, N. 1993. Flora of North America. Oxford Univ. Press, New York. Vol. 1, Introductory Essays, 320 pp.; Vol. 2, Pteridophytes and Gymnosperms, 320 pp.

Myint, T. 1966. Revision of the genus *Stylisma* (Convolvulaceae). Brittonia 18:97–117.

Radford, A.E., H.E. Ahles, and C.R. Bell. 1968. Manual of the vascular flora of the Carolinas. Univ. North Carolina, Chapel Hill. 1183 pp.

Samejima, K, and J. Samejima. 1987. *Trillium* genus illustrated. Hokkaido Univ., Sapporo, Japan. 237 pp.

Small, J.K. 1933. Manual of the southeastern Flora. Publ. by the author, New York. 1554 pp.

Yates, H.O. 1977. Revision of *Uniola* L. and *Chasmanthium* Link (Poaceae). SouthW. Naturalist 29:17–30.